# TRANSACTIONS

*of the*

American Philosophical Society

*Held at Philadelphia for Promoting Useful Knowledge*

Volume 84 Part 4

# Antonius de Carlenis, O.P. Four Questions on the Subalternation of the Sciences

Steven J. Livesey

Library of Congress Catalog
Card Number-94-72415
International Standard Book Number 0-87169-844-7
US ISSN 0065-9746

# TABLE OF CONTENTS

# ACKNOWLEDGMENTS

It is my pleasure to acknowledge the many individuals and agencies that have contributed to the production of this work. My thanks go first to the librarians of the several collections whose manuscripts have been consulted in the project, but especially those in the Bodleian Library, where much of the background work was completed. At the Vatican Film Library at Saint Louis University, Dr. Charles J. Ermatinger and his staff provided microfilms of many manuscripts that I would otherwise have been unable to inspect. Father William A. Wallace read the manuscript and kindly offered several suggestions and corrections, especially in the translation. At the University of Oklahoma, Professor Marcia Goodman and the staff of the History of Science Collections helped procure reference works for the introduction and critical apparatus.

Several agencies also supported the project. The initial transcription of the Bodleian and Trivulziana manuscripts was supported in part by grants in aid from the Fulbright Foundation of the United Kingdom and the National Science Foundation under grant number INT-8811869. Research at the Newberry Library and the Vatican Film Library was supported by travel grants from the Graduate Research Council at the University of Oklahoma and the Andrew W. Mellon Fellowship Program. The final preparation of the edition and the introduction was funded in part by the Oklahoma Foundation for the Humanities and the Southwestern Bell Foundation Fellowship Program.

Finally, my special thanks go to my wife, Nina, and my children, Daniel and Elizabeth, whose patience, understanding, and encouragement were indispensable.

Steven J. Livesey

Norman
June, 1993

# INTRODUCTION

The fifteenth century, it has been said, was one of the most perplexing and peculiar of recorded history. At the boundary between the modern world and the one it supplanted, this century contained much of both, so much in fact, that it is often difficult, perhaps impossible, for the historian to separate the two worldviews. Thus we commonly speak of continuity and contrast, of evolution and revolution being embodied in the fifteenth century.

It is perhaps in science, as Thomas Goldstein suggested a decade ago,[1] that this ambiguity is most striking. Certainly if one looks to the accounts of the history of science in the fifteenth century by previous historians, the confusion is more than apparent. For Pierre Duhem, premodern science consisted of twin peaks, one in the thirteenth and fourteenth centuries, the other in the sixteenth and seventeenth, when many of the earlier achievements were recalled and extended. Between them, philosophers of the fifteenth century merely mimicked or distorted and conflated the Parisian positions of the preceding century. Duhem's *bête noire* was frequently Paul of Venice, whose attempt to establish a middle path between Parisian terminism and Averroism established him as a "wretched philosopher" who was "not afraid of contradicting himself."[2] Following Duhem, Lynn Thorndike concluded that "scientifically the fifteenth century strikes us as distinctly inferior to the fourteenth, with the exception perhaps of certain fields such as surgery and anatomy."[3] And in summing up his work on Giovanni Marliani, Marshall Clagett observed that while Marliani was superior to most of his contemporaries, from the perspective of modern science he was inferior to his predecessors at Oxford and Paris. "This conclusion," he remarked, "seems to bear out the pessimistic judgment by both Duhem and Thorndike of the fifteenth century in comparison with that of the fourteenth."[4]

Other historians have been more sanguine, particularly in treatments of scientific method. In his celebrated work on the school of Padua, John Herman Randall, for example, argued that the Renaissance did not always entail a movement away from Aristotle and toward Plato, a repudiation of the scholastic tradition.

Rather, in Italy, and especially at Padua, Bologna, and Pavia, Aristotelianism was still a vital source of ideas: "What Paris had been in the thirteenth century," Randall suggested, "what Oxford and Paris together had been in the fourteenth, Padua became in the fifteenth: the center in which ideas from all Europe were combined into an organized and cumulative body of knowledge."[5] Among others, Randall's sources for investigation included the scourge of Duhem's story, Paul of Venice, whose order had sent him to Oxford in 1390, remaining there for three years before returning to northern Italy. Randall's central thesis that Paduan discussions of method produced a direct effect on Galileo has been largely rejected, yet his underlying emphasis on fifteenth- and sixteenth-century developments in Aristotelian philosophy found several supporters, including Charles Schmitt,[6] Cesare Vasoli,[7] and more recently William Wallace.[8]

Randall and his successors laid particular emphasis on the Paduan development of the two-fold method of resolution and composition and its relationship to *scientia propter quid* and *quia*, developed in the thirteenth century by Grosseteste and his followers and ultimately founded upon Aristotle's discussion in the *Posterior Analytics*. But this notion of scientific methodology was more fundamentally a part of Aristotle's discussion of scientific disciplines and organization, and included an adjunct discussion of the so-called *scientiae mediae*, whose medieval development also began with Grosseteste and was continued by many of the scholars Randall considered. In particular, Aristotle and his scholastic readers realized that while the disciplines of mathematics and natural philosophy were for the most part distinct, there were some disciplines whose subject matters and techniques crossed these otherwise distinct boundaries and were the sole exceptions to Aristotle's general prohibition against *metabasis*, that is, transdisciplinary work.[9]

This aspect of fifteenth-century methodological discussions largely eluded Randall, only to be taken up in part by Walter Laird in a pioneering dissertation. According to Laird, Paul of Venice relied heavily on the subalternation theory of Aegidius Romanus, but on occasion developed positions taken over from Robert Grosseteste. Paul's significance lies in the firm consolidation of the medieval tradition of subalternation theory. The extensive circulation of his *Expositio in libros Posteriorum Aristotelis* suggests that Paul was also instrumental in passing that tradition on to later scholars, although as Laird points out, this has never been demonstrated fully.[10]

It would go beyond the constraints of this essay to trace that influence comprehensively in the fifteenth century. Rather, I propose in part to examine the theory of subalternation as it was developed by one of Paul's readers shortly before mid-century, the Dominican friar and subsequently archbishop of Amalfi, Antonius de Carlenis de Neapoli.[11] According to his seventeenth-century biographers, Antonius was born of a noble family in Monte Aquilo, the mountainous region east of Cassino. Before entering the Dominican Order, he was a master of arts at Bologna, where in 1406–07 and again in 1407–08 he is reported to have taught Rhetoric.[12] Although he was present at the Council of Pisa (25 March–7 August 1409), perhaps as the secretary to the university's representative, Meersseman is probably correct in arguing that the seventeenth-century reports that Antonius impressed Alexander V by the strength of his disputations rests on a confusion of names.[13] At some later time, Antonius entered the Order, probably at the convent of S. Domenico in Bologna, where he is listed among the bachelors of theology in 1439.[14] His tenure as bachelor may have lasted until 1442, a date based largely on his possible attendance at the Council of Florence, which may have provided grist for his commentary on the *Sentences*.[15] In any event, in 1447, he is referred to as 'magister in sacra pagina' at the convent of S. Domenico in Naples, where at the same time he served as vicar and prior.[16] But in 1448, his ecclesiastical career took a new turn. It seems that Andreas de Pagliari, archbishop of Amalphi, had run afoul of both his flock and the pope—one account accused him of holding a concubine, another (probably earlier and more salubrious) account merely suggested that he was tainted by dishonest customs—, and that already in 1433 he had been accused by Eugenius IV. On 25 July 1448, Antonius de Carlenis, who was reported to have been on friendly terms with the archbishop, was appointed coadiutor, although Meersseman speculates that Antonius may have exercised some authority behind the scenes even before this. Perhaps fortunately, Andreas died 26 July 1449, and Antonius was consecrated archbishop on 11 August.[17]

However significant this promotion may seem to be, we must not forget that metropolitan rank in southern Italy was a pale shadow of its counterparts in northern Italy, to say nothing of northern Europe. Amalfi was a relatively small see, with only four suffragans. Of the eighteen metropolitan sees in southern Italy in the fifteenth century, only one—Rossano—had a smaller assessment for common services payable to the archbishop.

Unlike its neighbor, Naples, it did not enjoy the geographical eminence of a territorial lordship, and its economy had been in decline since the fourteenth century.[18] As a member of the regular clergy, Antonius seems to have accepted such a meager position when other seculars preferred wealthier or more prestigious sees.

Yet despite these modest resources, Antonius seems to have made the best of his situation. Considering his predecessor's problems, Antonius was perhaps an appropriate as well as a willing candidate, for in the bull in which he was appointed coadiutor, Antonius was described as having a "zeal for religion, knowledge of letters, purity of life, honesty of habits, the provision for spiritual things and caution in temporalities, and other merits of uprightness and virtue."[19] His personal life is virtually unknown; he did construct a chapel to St. Thomas in the cathedral, notable perhaps because of his adherence to the "Angelic Doctor" 's teaching. When Antonius died on 23 May 1460, he was buried here in the chapel.[20]

The decade before his elevation to the see in Amalfi seems to have been intellectually a productive one for Antonius. We have already mentioned that his commentary on the *Sentences* was produced at Bologna, probably during the academic year 1439–40. But internal evidence from the *Questiones in libros I–II Analyticorum Posteriorum Aristotelis* suggests that it was produced after the *Sentences* commentary. In Book I, question 22, in his discussion of the formal requirements for subalternate sciences, Antonius refers to what he calls "that question which is famous in theology, . . . that [theology] is subalternated to the science of the Blessed." He notes that Thomas and Scotus stood on opposite sides of the issue, and that Aegidius Romanus had specified three conclusions bearing on this issue. "But," he says, "concerning this we have spoken in the first book of the *Sentences*."[21]

Beyond this, precisely where and when Antonius wrote his commentary on the *Analytics* remains unknown. Antonius's sources in the commentary were for the most part well known and widely held. The two notable exceptions, Johannes Brito and Jacobus Forlivio, cannot be placed in the Dominican libraries of either Naples or Bologna, although the Augustinian library of San Giacomo in Bologna seems to have held an impressive number of Antonius's sources, including the latter.[22] In the absence of additional information, one must suspend judgment on the location and precise date of origin for Antonius's commentary.

The sources upon which Antonius drew, in fact, betray the traditional nature of both the *Questiones in IV libros Sententiarum* and the *Questiones in libros I–II Analyticorum Posteriorum Aristotelis*. The authors that Antonius cites in the four questions edited here are listed in descending order of their frequency, a list which probably would not change appreciably if his citations throughout the commentaries were tabulated:

| *Questiones in libros* *Sententiarum* Prol. q. 1–2 | | | *Questiones in libros* *Analyticorum Posteriorum* I, q. 17, 22. | | |
|---|---|---|---|---|---|
| Hervé Natalis | 17 | (17.5%) | Aristotle | 45 | (57.0%) |
| Aquinas | 13 | (13.4%) | Aquinas | 12 | (15.2%) |
| Duns Scotus | 10 | (10.3%) | Aegidius Romanus | 9 | (11.4%) |
| Landulfus Caracciolo | 9 | (9.3%) | Paul of Venice | 5 | (6.3%) |
| Franciscus de | | | Averroes | 3 | (3.8%) |
| Mayronis | 8 | (8.2%) | Landulfus Caracciolo | 2 | (2.5%) |
| Aristotle | 7 | (7.2%) | Duns Scotus | 1 | (1.3%) |
| Petrus Aureoli | 6 | (6.2%) | Walter Burley | 1 | (1.3%) |
| Aegidius Romanus | 5 | (5.2%) | Franciscus de | | |
| Durand de S. Pourçain | 4 | (4.1%) | Mayronis | 1 | (1.3%) |
| Johannes de Neapoli | 4 | (4.1%) | Totals | 79 | (100.0%) |
| S. Biblia | 4 | (4.1%) | | | |
| Girardus de Bononia | 3 | (3.1%) | | | |
| Godefroi de Fontibus | 2 | (2.1%) | | | |
| Henricus Gandavensis | 2 | (2.1%) | | | |
| S. Augustine | 1 | (1.0%) | | | |
| S. Jerome | 1 | (1.0%) | | | |
| Hermanus de Maio(?) | 1 | (1.0%) | | | |
| Totals | 97 | (100.0%) | | | |

The most noteworthy observation to be made from the list is that with few exceptions, Antonius's post-Patristic sources were Parisian masters, and only Paul of Venice could be considered a recent source.[23] The other is that despite Antonius's reputation as a Thomist, frequently his information about Thomas's position is filtered through Aquinas's followers or pupils, especially Hervé Natalis and Aegidius Romanus. In fact, in the first two questions from the prologue to the commentary on the *Sentences*, these two sources combined to produce nearly double the number of citations from Aquinas himself.

This conventionality extends to the format of both commentaries—and to his commentary on the *Metaphysics* as well[24]—, which follows the model of late thirteenth- and early fourteenth-century commentaries. Antonius begins with the statement of the question and the initial position, several (on occasion, as many as twenty) arguments in its support, the *sed contra*, which includes in the commentary on the *Posterior Analytics lemmata* from the text, and the solution to the question, divided into two sections: first the *notanda*, frequently four initial observations on the problem, then conclusions—again frequently four in number[25]—that form parts of the syllogistic solution to the question. Finally, Antonius returns to resolve the principal arguments to the question. In both its sources and its format, these are commentaries that might have been composed a hundred years earlier.

• • • • •

The few modern readers who have analyzed his works have usually regarded Antonius as a Thomist. But as early as the fifteenth century, there were doubts about the purity of his Thomism; his Dominican confrère, Dominicus de Flandria (d. 1479), accused him of Scotist tendencies among other things.[26] Given these conflicting perceptions of his thought, Antonius may merit additional investigation, as Tuszyńska has suggested.[27]

Quite apart from the issues of existence and essence, the analogy of being, or discussions of the unity of substantial form, the question of the subalternation of the sciences and the issue of the scientific status of theology can be regarded as *topoi* for evaluating congruence between Thomas and his reputed followers, for it was in large part Thomas's position on these matters that stimulated continued discussion through the end of the Middle Ages. The "questio famosa" to which Antonius referred in his commentary on the *Analytics* reflects that tradition and is the subject of the opening two questions of his commentary on the *Sentences*, namely whether theology is a science, and if so, whether it is a subalternate or subalternating science.

In the first question, Antonius lays out the essential Thomistic contention that theology is a science and the denial of that position by Scotus and Peter Aureol. He observes that for Thomas, theology is both *scientia* and *sapientia*, while for his opponents, it is more the latter than the former. According to them, even if some of the premises of theology are known, others are merely believed, and according to Aristotle, the conclusion derived from believed and known premises will be merely contingent, and hence not the object of science. To answer his (and Thomas's) critics, Antonius borrows from Hervé Natalis the distinction between *quia* and *propter quid* acceptance of principles in the subalternate science. Sometimes theology proceeds from principles that are known per se *quantum ad quia est*, and thereby is a science properly speaking. At other times, theology does not have the evidence of its own principles, except insofar as they are proven in the subalternating science *quantum ad propter quid est*. But the crucial contention that Antonius borrows from Hervé (and Thomas) is that in order to prove something in the subalternate science, it does not suffice to argue from merely *believed* things any more than the perspectivist merely believes his first principles. The distinction between theology as science and wisdom is a false dichotomy, since it rests on different senses of the criteria for science.[28]

In his second *notandum*, Antonius approaches the same issue from a slightly different perspective: he notes that Scotus, Peter Aureol, and Franciscus de Mayronis all rejected Thomas's contention that the articles of faith are the principles of theology and that they constitute the nexus of subalternation between our theology and that of the blessed. In response, Antonius once again cites Hervé, this time supported by his compatriot, Johannes de Neapoli, that theology is a science *largo modo*, that is, when considered a subalternate science. Moreover, if one looks only at the conclusions of theology, it is scientific in a strict sense, since the conclusions follow necessarily from the principles.[29]

In fact, Antonius suggests five ways by which theology may be scientific: first as an enterprise parallel to metaphysics; second insofar as its evidence exceeds our capacities; third, insofar as it is possible for it to know its own principles; fourth, broadly speaking, since, as Antonius says, "We (that is, the theologians) know many things about the articles of faith that the simple believer does not"; and finally, as a science of consequences, since the conclusions of theology follow necessarily from its own principles.[30]

Antonius also mentions Henry of Ghent's theory of a *medium lumen* between the understanding of faith and that of a face-to-face encounter of the blessed, but he says that he will pass over the theory because Thomas does not speak of it.[31] And notwithstanding his discussion in the second *notandum*, Antonius says in his fourth *notandum* that theology is more than merely a science of consequences, calling once again upon Hervé for support. Beyond its own internal logic, theology is a science of consequences that produces understanding of its subject matter, and this distinction between *logica docens* and *logica utens* is, he says, consistent with Aristotle's discussion of science in the *Posterior Analytics* and the *Metaphysics*.[32]

From these four preliminary arguments, Antonius derives three conclusions. First, the theological habit is not a habit of

faith, for the substance of all four *notanda* was that theology is in
a certain sense a scientific habit. Second, Antonius argues that
theology must be a discursive habit, and his proof is essentially
one of eliminating other possibilities: it cannot be an innate habit
or a habit of assent, since as Peter Aureol had argued, if it were
simply a matter of faith, the theologian would have spent many
fruitless years acquiring theological understanding. Finally, Antonius argues that theology must be an acquired, not an infused
habit, at least in the sense that Thomas had suggested, in which
conclusions are proven from principles.[33]

Although it had already occupied a significant portion of the
first question, Antonius focuses more specifically on the issue of
subalternation in his second question. Once again, he draws on
Hervé Natalis for much of his information. He refers, for example, to a dual sense of subalternation, *ex parte scibilis* and *ex parte
scientie*, and following Hervé argues that the former cannot apply
to the case of theology, since the same thing under the same formal *ratio* is present in both our theology and the knowledge of the
blessed. Nor can one say strictly that the latter applies either.
Only when one takes science in a general way as a cognitive habit
commonly speaking can one speak of a subalternating relationship between the science of the blessed and our theology, because a habit of belief is subordinate to an evident habit.[34]

In his second *notandum*, Antonius refers to objections to Aquinas's theory made by Scotus, Peter Aureol, and Landulfus Caracciolo, who rejected the suggestion that subalternation theory
could apply to the science of theology. Antonius borrows his response from Aegidius Romanus, who agreed that while none of
the traditional definitions of subalternation sufficed for this special case, one could speak of a similarity between subalternation
of the human sciences and the relationship binding our theology
to that of the blessed. Strictly speaking, our theology and that
of the blessed do not differ in the nobility of the object—both
are concerned with God—, but they apprehend him more excel-
lently than we. Furthermore, the *propter quid/quia* relationship
between the subalternating and subalternate sciences does not
apply strictly speaking, because God is the cause of all things
immediately or mediately; yet the principles that we use in theology are less than transparent, while they are evident to the
blessed. And finally, just as the astronomer and the navigator
know the positions of the stars in a subtle or crude way respec-
tively, so the blessed and the viator labor under different degrees
of knowledge.[35]

Antonius also raises the objection made first by Peter Aureol, that under Aristotle's theory of subalternation, no one can have an awareness of the subalternate science without also having knowledge of the subalternating science, clearly an inconvenient position in the case of our theology and that of the blessed. His response follows that of Girardus de Bononia, who argued that one can also acquire knowledge of principles in the subalternate science through the senses, memory, and experience.[36]

In keeping with his reliance upon Hervé Natalis, in his third *notandum* Antonius criticizes an argument made by Durand de St. Pourçain. Following the *via Scoti*, Durand had argued that the subalternate and subalternating sciences ought to be always compossible in the same subject, since one is confirmed by the other, as in the case of perspective and geometry. But this is not so for the case of our theology and that of the blessed, and consequently they cannot be related as subalternate and subalternating sciences.[37] Citing Hervé, Antonius responds by denying the assumption, since in subalternation *ex parte scientis*, especially where there is a dependence of faith on knowledge, there is no compossibility, since it would necessitate clarity and diversity with respect to the same object. But this, as Antonius argued in the first *notandum*, is the limited sense under which one can view theology as a subalternate science, and hence only the strong comparison with perspective and geometry or the other pairs of human subalternate sciences should not be made.[38]

Finally, Antonius returns to the problem of consistency in Thomas's theory. Sometimes Thomas had said that neither faith nor theology were concerned with known—apart from believed—things, while at other times he spoke of theology as a subalternate science. With Hervé, Antonius sees no contradiction, at least when one affirms the two senses of subalternation discussed in questions one and two of his prologue: strictly speaking, theology is not a science, but in a more general sense, when one takes the definition of subalternation as the dependence of principles in one science on those in another, theology is a subalternate science, and hence truly scientific. And once again, following Hervé and Aegidius, Antonius asserts that what Thomas was really speaking of was the similitude of subalternation in human sciences and the relationship between our theology and the knowledge of the blessed.[39]

The emphasis on similitude permeates the remaining conclusions Antonius proves in question two. Theology is not subalternated to metaphysics, nor does it subalternate to itself other human sciences, despite its appellation as queen of the sciences.[40] Subalternation obtains only in the limited sense of the dependence of our theology on the knowledge of the blessed.[41]

This brief summary of Antonius's defense of the Thomistic theory of subalternation reaffirms the suggestion that this is a Thomism filtered through subsequent sources, especially Hervé and Aegidius. In fact, one may question the degree to which Antonius had scrutinized the works of either Thomas himself or his critics, even those he cites. There is nothing in these two questions to indicate that he appreciated Thomas's discussion of the formal/material aspects of science, upon which Thomas based his theory of subalternation, and which many subsequent readers of Thomas—among them several Scotists—criticized. On the other hand, some arguments made by the critics Antonius cites— including references to the criterion of the superadded condition that distinguishes the subalternate and subalternating subject, which became virtually ubiquitous in the fourteenth century— are overlooked in his analysis. There may be several reasons for these omissions. One is that in defending Thomas, Antonius responded only to the arguments he found in the sources he used. But more likely still is the suggestion that Antonius was most comprehensive when he relied on earlier defenses of Thomas, while his examination of subsequent critics—especially Peter Aureol—was more superficial. Thus Hervé and Aegidius provided both the structure of the analysis and the arguments used to substantiate it.[42]

Several years ago, Paul Kristeller observed several characteristics common to followers of schools of philosophy.

The philosopher who follows the authority of a master . . . is strongly inclined to understand and interpret the master with the help of his own powers and vision according to his own conception of the truth. He must clarify the ambiguities, obscurities, and inconsistencies he discovers in the text and thought of the master, must fill his gaps, and apply his principles to the solution of new problems as well as to the refutation of new rival positions not known or studied by the master.[43]

But if the ability to address new problems unknown to the master is a touchstone of excellence, Antonius likewise shows deficiencies as a Thomist. Among the arguments Antonius treated in these two questions, one in particular occupied his attention and that of many followers and opponents of Thomas's theory, namely the epistemological status of belief and knowledge as the basis for subalternation of our theology to that of the blessed. Antonius's discussion omits an argument that became somewhat popular as the fourteenth century progressed, and which indeed is somewhat characteristic of the century. In concise form, it appears in the first question to the prologue of Gregory of Rimini's commentary on the *Sentences*. As an argument confirming Thomas's theory, Gregory suggests that

God could conserve the habit of perspective in the intellect of someone, while not conserving in it the habit of geometry. But in such an intellect, that perspective would truly be a subalternate science, and nevertheless it would not have understanding of its own principles, but only faith.[44]

In response to this hypothetical (and clearly self-serving) argument, Gregory notes

I say first that the same argument could be made of a science [that was] not subalternate, supposing that the intellect of the principles be distinguished from the habit of the conclusion, and so equally they could have said absolutely that theology is a science without adding that it was subalternate (which nevertheless they do not). Second, I say that if God should conserve some perspective (which now *de facto* is science) without [conserving] geometry, still it would be science, having suitably arisen in the same acts in which it is now, other things remaining the same. However, it does not follow from this that perspective which is acquired while geometry was lacking would be science.[45]

Gregory thus attacks the hypothetical argument (which he believes entails the Thomist theory of subalternation of theology) first by suggesting that Thomas's distinction of science and the restriction of only one sense to theology is a ruse, and that the very same argument could be made without resorting to subalternation, something that Thomas would be reluctant to do. More interesting, however, is his second argument, in which Gregory would distinguish between conclusions acquired before God ceased to conserve geometry in the mind of the knower and those acquired afterward. The former would remain scientific even after the knower had no awareness of geometry; the latter would never be considered scientific, and since under normal circumstances it is the latter which pertains in the strongest sense to the case of our theology, Thomas's argument would be without merit.

One might expect Antonius to respond to this argument which cuts to the central issue of Thomas's theory. One might even suggest that it was appropriate to his discussion in the second *notandum* of question 2, in which he attacked Aureol's position on the inseparability of the subalternating and subalternate sciences.[46] Yet he does not. It is, of course, possible that he was not aware of it, despite its popularity during the fourteenth century. He does not cite Gregory of Rimini in these preliminary matters on subalternation, and I have found no indication that he used Gregory's commentary elsewhere in his work. The same argument was available to him substantially, however, in Peter Aureol's *Scriptum*, where it is cited as a potential argument for the Thomist theory of subalternation, but one that Thomas never made explicitly.[47]

Antonius was not alone among followers of Aquinas in neglecting to respond to this argument. Johannes Capreolus, the "prince of Thomists" at least among theologians and otherwise a perceptive reader of both Peter Aureol and Gregory of Rimini, fails to mention the argument at all in his discussion of subalternation.[48] Nor did Denis the Carthusian, whose commentary on the *Sentences* frequently served as an encyclopedic source for medieval arguments.[49] In England, the same is true of Thomas Claxton, who undertook a long discussion of the scientific status of theology that cited Gregory of Rimini among other fourteenth-century Parisian masters.[50] While Antonius's omission probably resulted from his secondary knowledge of Peter Aureol and his narrow investigation of fourteenth-century materials, that omission seems to have been quite common among fifteenth-century Thomists.

• • • • •

When one turns to the commentary on the *Posterior Analytics*, however, Antonius becomes more complete and comprehensive in his discussion of subalternation. In Book I, question 17, Antonius asks whether demonstration can proceed from extraneous premises. This was a question that both Aristotle and his medieval commentators considered preliminary to the issue of subalternation, for in Book I, chapter 7, Aristotle had prohibited the illegitimate transference from one domain to another in the process of demonstration. There and again in I. 9 and I.13, Aristotle recognized that certain demonstrations did not follow, strictly speaking, this autonomy of proof, and he explicitly made the subalternate sciences the sole exceptions to his prohibition of *metabasis.*[51] Thus the issue of disciplinary autonomy provides the context for the more specific discussion of subalternation of the sciences.

Antonius introduces this connection at the beginning of his discussion in question 17: what did Aristotle mean by requiring that demonstrations proceed from the same genus? From Thomas—and ultimately from Aristotle himself—, he introduces the distinction between *idem genus secundum quid* and *quodammodo.* According to Antonius, Aristotle's prohibition of *metabasis* was intended to cover those sciences whose subjects were entirely distinct. But beyond this, Antonius wishes to steer an intermediate course between strict identity of subjects that would confine scientific demonstration unnecessarily and unbridled appeals to similarity of subject genera—like Brisso's analogical techniques condemned by Aristotle—that fail to distinguish between common and proper principles.[52]

Antonius also discusses a minor exegetical issue very briefly: if Aristotle proposed that the three things involved in demonstration were the subject, its properties, and the axioms, and that these defined the genus of demonstration, where should one include the premises of the proof? Aegidius had argued that while Aristotle did not mention them explicitly, they ought to be understood under the category of axioms. Antonius notes that Paul of Venice had objected to this interpretation, arguing that if the premises and middle terms were reducible to axioms [*dignitates*], Aristotle would not have mentioned subject and properties, since these also are reducible. But Antonius dismisses this objection as too focused on literal interpretations [*verbales*], and so does not address it.[53]

According to his third *notandum*, sciences must be based on proper, not common principles, and for further specification of what Aristotle meant by extraneous principles, he relies upon Paul of Venice. Something can be said to be extraneous in two ways: either because there is no internal agreement (for example, that two cubes make a cube is extraneous to triangle) or because one could find something else where the thing would fit more appropriately (in the sense that genus and species are extraneous, and common principles are extraneous to the most specific species). In this second sense, although the genus is specified in the definition of the species and things are predicated universally of it, such a proposition would not be non-extraneous (to use the double negative of the text). The result is that a common thing is from different perspectives both inherent and extraneous.[54] This ambiguity is, in fact, one of the reasons Aristotle had been so pre-occupied with the appropriate use of common principles.

Antonius summarizes his discussion to this point by drawing two conclusions. First, demonstrations must be from proper, not common principles. Second, even when one speaks of the sub-alternating science descending into the subalternate science, such sciences never proceed from extraneous principles or prin-ciples that are common precisely in the aspect by which subal-ternation takes place. However, Antonius immediately suggests that on occasion this may happen, but he refers discussion of this problem to the special question on subalternation.[55]

Finally, in the fourth *notandum*, Antonius restates the distinc-tion between common and proper principles that Aristotle had made in *Posterior Analytics* I.10. Within the specific genus, both the existence and the definition of the basic entities are assumed, and common principles are made specific to the genus by a pro-cess of analogy. All subsequent matters are proven.[56]

From these assumptions, Antonius's conclusions follow imme-diately: sciences proceed from proper, never common principles. And, as Aristotle had required, demonstration never descends from one genus into another. Otherwise, Antonius says, demon-stration would not proceed from proper principles, but rather by means of a descent from one genus to another subalternated to it in the relationship of a species to a genus.[57]

In question 17, Antonius had said little about subalternation beyond making the tie to the prohibition of *metabasis*. However, this final argument leaves the impression that he saw subalter-nation in part as describing a genus-species relationship between sciences. In question 22, Antonius presents a more detailed pic-ture of subalternation, and in his first *notandum*, his initial re-marks preserve this impression. There are two ways, he says, by which sciences are subalternated, either because their subjects are related as genus and species, or because those subjects com-pare as formal and material. While he calls upon St. Thomas for support, he clearly has not reproduced the master's text scrupu-lously, for Thomas was emphatic in asserting that while both these criteria define ways by which one science can be said to be under another [*una scientia esse sub altera*], only the second defines subalternation. Antonius's subsequent quotations from Thomas suggest that this should have been apparent, but the imprecise definition in his introductory remarks makes this unclear.[58] In the second *notandum*, for example, Antonius seems to intend an elaboration of Thomas's material/formal theory, but the example he presents is rather puzzling. Pure mathematics, he says, concerns a subject which is more abstract and formal, while the subalternate sciences, like perspective and arithmetic, concern subjects which are contracted and pertain to matter.[59] Aside from the problem of clarifying his position on the genus-species relationship in subalternation, Antonius seems to make the novel suggestion that arithmetic is subalternated to undifferentiated mathematics, and that perspective is as well.[60]

The issue of material and formal aspects of a subject leads Antonius to report and then criticize Paul of Venice's theory of mathematical abstraction. In contrast to Aegidius Romanus and Aquinas, who held that the mathematician abstracts only from qualitative matter, Paul argued that the mathematician abstracts from all substance. According to Paul, while the mathematician does not abstract from intelligible matter in the sense of continuous or discrete quantity, he does abstract from it in the sense of prime matter considered without form and from sensible matter, which is prime matter considered with sensible quantitative form. Although quantity usually is held to inhere in substance primarily, and although it usually presupposes substance in the natural order, it does not include substance in its quidditative concept, and so the mathematician can abstract quantity from all substance. As a proof text for this assertion, Paul cites Aristotle's remark in *Physics* II.2 that geometrical things inhere in a subject, yet they are understood without the subject.[61]

In response, Antonius says that since subject falls under the definition of its own accident, in the case of quantity, if it is defined quidditatively, subject ought to be posited of it as well. But then Antonius backs away from his own argument: lest it be assumed, he says, "that the geometer must know the concept of substance when he wishes to define the concept of quantity, I say that the geometer does not define the quiddity of quantity, but <rather> the proportions or measurements and such things under which quantity occurs."[62] And so while he begins by opposing Paul's somewhat unusual definition of abstraction, `he succeeds in something less than full refutation.

As to the criteria for subalternation, Antonius argues that merely having one subject under another is insufficient, a categorical statement that resolves his earlier ambiguities. According to Antonius, two conditions are necessary: one subject must be under another in such a way that the subject of the subalternating science is contracted [*sit contractum*] by some accidental *differentia*, and the superior science must speak *propter quid* of the same thing that the inferior science speaks only *quia*.[63] Antonius's use of *contractum* presents an intriguing possibility. While his sources for the theory of subalternation had been Aegidius and St. Thomas, this is an element that he may have taken from Paul alone, and if Laird is correct, one that indirectly connects Antonius with the earlier theory of Robert Grosseteste.[64]

Finally, among his conclusions in this question, Antonius refers in passing to another central issue in medieval subalternation theory, namely the partial subalternation of one science to another. There is nothing, he says, to prevent the same science from being both a subalternate and subalternating science, at least with respect to different parts. The science of the rainbow, for example, is subalternated to both perspective and natural philosophy, and (citing Aegidius and Paul once again) he says that while surgery is not wholly subalternated to geometry, nevertheless it is in part.[65]

· · · · ·

It may be helpful in assessing Antonius's position on subalternation of the sciences to compare his treatment with that of his fifteenth-century confrère, Dominicus de Flandria. As Meersseman has shown, Dominicus strongly criticized Antonius's adherence to Thomas, attacking the archbishop's commentary on the *Metaphysics* in his own discussion of the work. But did this disapproval extend also to the logical works, and to the discussion of subalternation?

While Dominicus seems not to have produced a commentary on the *Sentences*, his commentary on the *Posterior Analytics*, produced in 1475 at Florence, contains materials on both *metabasis* and the subalternation of the sciences. As one might expect, there is some similarity between the two treatments, in part because both commentators were following Aristotle through Thomas. In fact, Dominicus constructs his commentary even more carefully as an exposition on Aquinas, proceeding lecture-by-lecture through the Angelic Doctor's work. Thus, for example, under lecture 15, we find two questions, "Utrum in demonstrationibus possit fieri discursus de genere in genus" and "Utrum extrema demonstrationis et eius medium sint eiusdem generis."[66]

With respect to the first question, Dominicus responds in the usual way, by recognizing that under most circumstances, there are strict limitations placed upon the freedom to choose the terms and premises of demonstration. The underlying concern is, as we have seen, that without such strictures, demonstration would be accidental, not essential. But this discussion leads Dominicus to inquire about the nature of the subject genus of demonstration, in which he draws upon the familiar formulae of Aegidius Romanus and Aquinas, that the genus might be absolute or with respect to something else. Finally, he observes that

. . . genus might be understood in two senses: in one way as a natural genus; in another way, as a logical genus. The latter can be understood in two senses: as the genus of predicables, or as the subject genus or the genus of things known, to which some proper attribute can be demonstrated. It is in the final sense that genus is understood here in this context, and not in the first two senses.[67]

Both the context and the content recall Dominicus's remarks in his commentary on the *Metaphysics*, in which he refers to Antonius's definition of subject genus:

Aliqui, ut Antonius de Neapoli, quidam Archiepiscopus, Q. 1 in sua Metaphysica, in primo notando, et alii similes distinguunt inter subiectum attributionis et subiectum adequationis, vocantes subiectum adequationis esse subiectum principale scientiae per quod scientiae denominationem accipiunt et distinguuntur. Subiectum autem attributionis dicunt esse principaliores partes consideratae in illa scientia, sicut sunt substantiae separatae in hac scientia. Sed isti, cum reverentia, propriam vocem ignorant.[68]

However, at the passage in the *Analytics* commentary, Dominicus engaged in no polemics. Instead, he concludes by observing that the principles of demonstration are either proper to their disparate genera (and hence no transference is possible) or common, and if the latter remain uncontracted by an appropriate subject matter, they are common to several sciences. If Dominicus knew Antonius's commentaries on the *Sentences* or the *Posterior Analytics*, he chose not to take issue with them in this context.[69]

In his comments on lecture 25, Dominicus considers two questions on the subalternation of the sciences: "Utrum demonstratio propter quid et quia differant in diversis scientiis," and "Utrum philosophia naturalis subalternetur perspectivae, et perspectiva geometriae."[70]

In his discussion of the first question, Dominicus observes with Aristotle and Thomas that the *quia/propter quid* relationship occurs within those sciences that are related as subalternate and subalternating, but that on occasion, it can be found among sciences that are not strictly speaking related in this way, and Dominicus cites the well-known example of medicine and geometry and the healing of wounds.[71] This brings Dominicus to review Thomas's position on the definition of or criteria for subalternation, material we have already seen in Antonius's presentation. Like Thomas and Antonius (in the commentary on the *Analytics*, but not the *Sentences*), Dominicus acknowledges two senses of subalternation: when the two subject genera are related as genus and species, and when they are related materially and formally. The latter is the proper understanding of subalternation as Aristotle and Thomas had maintained. Drawing upon Aegidius Romanus and Paul of Venice, Dominicus introduces the discussion of the virtual univocity of the two disciplines, in which the distinction nevertheless arises because the superior science examines the subject abstractly and through causes, while the inferior science considers its subject only concretely, restricting its examination to consideration of phenomena.[72]

As to the specific relationship between natural philosophy and perspective, Dominicus also adopts the common position of partial subalternation, which he says he has taken from Aegidius Romanus, from Thomas, and other doctors of natural philosophy. Thus, while the whole of natural philosophy is not subalternated to perspective in its entirety, one can find a part—viz. the science of the rainbow—in which this is the case.[73] The remainder of his discussion in this question is devoted to a survey of the various positions on the definition of subalternation, which have been taken from Aegidius Romanus (and augmented by Paul of Venice),[74] Albert the Great,[75] and Albert of Saxony.[76]

Three conclusions follow from this brief examination of Dominicus's discussion. The first is that he seems to have felt an obligation to augment Thomas's discussion of *metabasis* and subalternation by presenting synopses of the prominent views since Aquinas's time. Although his sources are not identical with those used by Antonius, there is considerable similarity, and both seem to have given less consideration to late fourteenth- and fifteenth-century sources than to those from the period immediately following Thomas. Dominicus's use of sources can in some sense be considered even more encyclopedic than Antonius's, since on some occasions—as in the response to the second question in *lectio* XXV—the distinctions drawn from his sources are never fully integrated into the question at hand.

On the other hand, it seems that at least in these passages, Antonius was a more argumentative purveyor of Thomist positions on *metabasis* and subalternation. Thus, for example, he engages in the traditional arguments against Scotus and Peter Aureol, but he also finds it necessary to extend his criticism to Paul of Venice on occasion, weak though it may be. By contrast, Dominicus's treatment of the subject is a more assertoric or descriptive one, and as a result, Antonius's approach to the material appears to be somewhat more complex by comparison.

Finally, while Dominicus refers occasionally to the *schola Thomistarum*, there is no evidence to suggest that Antonius was intended by this designation. As we have seen, when the context might have suggested it, references to Antonius's commentaries are noticeably absent.[77] If Dominicus was aware of Antonius's discussion of subalternation and *metabasis*—and there is no evidence to suggest that he was—he did not include it in his own treatment of the topics.

• • • • •

The picture that emerges from Antonius's commentaries on the *Sentences* and the *Posterior Analytics* is that he was both eclectic in his choice of sources and somewhat idiosyncratic in his use of them. It is clear that his fundamental position was Thomistic, but his Thomism was filtered through the subsequent tradition of Aegidius Romanus and Hervé Natalis, among others. On the other hand, he was not reluctant to use materials outside the Thomist tradition as well, and there can be no better example of this than his use of Paul of Venice. Although Antonius cited Paul on occasion as an alternative view regarding the issue of subalternation, on virtually every occasion, he concludes by either agreeing with Paul, issuing no opinion, or providing only a mild or inconclusive objection. Although his acquaintance with the literature of subalternation was not as broad as others in the Thomist tradition, his comments reflect the consolidation—and indeed augmentation and alteration—of the Thomist position that can be seen in other commentators of the fifteenth century.

But if Antonius occasionally strayed from the Thomist fold either through ignorance, carelessness, or excessive reliance upon his sources, his fundamental position on subalternation was a conservative one. Beyond the several examples already examined, we may see this tendency in his response to one of the principal arguments of question 22. According to that argument, if the subalternate science operates on a contracted subject, it seems that either the subalternating and subalternate subjects were as genus and species, and hence reducible to the same subject, or the subalternate subject was dependent upon some accidental property. Antonius responds by citing Thomas, but then poses a hypothetical argument, asserting that 'visual line,' the subject of perspective, or 'sonorous number,' the subject of music, although accidental in being, can be the subject of a *per se* proposition that is scientific. In response, Antonius points out that this was a special case under different circumstances, and that those circumstances do not apply in the case of subalternation. Antonius denies that a *per accidens* aggregate can in any way become the subject of science, and in doing so seems to repudiate the tactic used frequently in the fourteenth century to expand the technique of subalternation.[78] As in Thomas and ultimately Aristotle, subalternation was merely a narrow exception to the general rule that prohibited *metabasis* in scientific demonstration.

Finally, where does this leave Antonius in relation to his contemporaries? It would, of course, be misleading to place him among the first-line philosophers of the fifteenth century. But similarly, it would be incorrect to assume that as a second-line scholar whose works had limited circulation, he has little historical significance. Until the rest of his work is studied more completely, it will be difficult to determine the full extent of this significance, but the limited investigations in this and Meersseman's earlier research point to an area of Antonius's methods that characterize him as an eclectic philosopher, a feature shared by many of his century.

In his fundamental study of Renaissance Aristotelianism, Charles Schmitt emphasized that eclecticism offered various strategies. On the one hand, Renaissance Aristotelians often accepted new developments, especially in science and formalistic disciplines, that provided superior doctrine to that of traditional Aristotelian sources. On the other hand, borrowing from other sources also offered the long-term strategy that one's own philosophy might be strengthened. And so one finds, particularly in the sixteenth century, the invocation of Plato, the Hermetic tradition, *prisca sapientia*, and a host of other sources to accomplish this goal.[79]

Antonius, it seems, was both too early chronologically and too distant philosophically to have employed these particular sources, yet he remained eclectic in another way, one he shared with many of his scholastic predecessors and contemporaries. This was one of the reasons his confrère, Dominicus de Flandria criticized him; his Thomism was not pure. The extensive use of earlier scholastic sources by fifteenth-century commentators[80] has sometimes been responsible for the negative assessment of the period discussed above (pp. vii–viii), especially Duhem's characterization of Paul of Venice. But Duhem certainly had it wrong: in his eagerness to look back to the fourteenth century, he foreclosed any possibility that the fifteenth-century eclectic collections may have provided a model for later eclectics and their incorporation of other philosophical traditions. As Schmitt clearly observed, the Western European Aristotelian tradition had never been a monolithic structure,[81] but between thirteenth- and fourteenth-century interpretations and those we usually associate with Renaissance Aristotelianism, there lay a distinct but important group of consolidators of the earlier tradition. Seen from this perspective, the significance of Antonius's work and that of his second-line contemporaries may lie less in the transmission of particular positions on philosophical issues than in contributions to techniques of eclectic Aristotelianism.

## Notes

1. Thomas Goldstein, *Dawn of Modern Science* (Boston: Houghton Mifflin 1980), 191.

2. This strain in Duhem's work can be seen in several places, but concerning Paul of Venice, see *Le Système du monde* (Paris: Hermann 1913–1959), vol. 10, p. 412; vol. 10, p. 422; see the translation by Roger Ariew, *Medieval Cosmology: Theories of Infinity, Place, Time, Void and the Plurality of Worlds* (Chicago: University of Chicago Press 1985), 291, 427 and my essay review in *Science in Context* 1(1987): 363–370 at 366–367.

3. Lynn Thorndike, *A History of Magic and Experimental Science* vol. 4 (New York: Columbia University Press 1934), 614. However, in the introductory chapter to *Science and Thought in the Fifteenth Century* (New York: Columbia University Press 1929), he gives a somewhat more positive view.

4. Marshall Clagett, *Giovanni Marliani and Late Medieval Physics* (New York: Columbia University Press 1941), 170.

5. John Herman Randall, Jr., *The School of Padua and the Emergence of Modern Science* (Padua: Editrice Antenore 1961), 24. Randall had occasion several years later to respond to his critics; see, for example, "Paduan Aristotelianism Reconsidered," in *Philosophy and Humanism. Renaissance Essays in Honor of Paul Oskar Kristeller*, ed. E. P. Mahoney (New York: Columbia University Press 1976), 275–282.

At the other end of the spectrum on this issue, Neal Gilbert's assessment of the fifteenth-century developments was that "research into fifteenth-century thought has all the difficulties but offers little of the excitement of research into medieval thought. . ."; see *Renaissance Concepts of Method* (New York: Columbia University Press 1960), 164.

6. See, for example, Schmitt, "Towards a Reassessment of Renaissance Aristotelianism," *History of Science* 11(1973): 159–193 esp. at 171ff; reprinted in *Studies in Renaissance Philosophy and Science* (London: Variorum 1981), and *A Critical Survey and Bibliography of Studies on Renaissance Aristotelianism (1958–1969)* (Padua: Ed. Antenore 1971).

7. Cesare Vasoli, "La cultura dei secoli XIV–XVI," in *Atti del primo convegno internazionale di ricognizione delle fonti per la storia della scienza Italiana: I secoli XIV–XVI*, ed. Carlo Maccagni (Firenze: G. Barbèra 1967), 31–105. See also Charles Schmitt's remarks about Vasoli's work [*A Critical Survey . . .* pp. 36–39].

8. William A. Wallace, "The Medieval Accomplishment in Mechanics and Optics," in *Prelude to Galileo. Essays on Medieval and Sixteenth-Century Sources of Galileo's Thought* (Dordrecht-Boston: Reidel 1981), 29–63, esp. at p. 44ff., where the fifteenth century is viewed as a period of transition.

9. For a more detailed discussion of Aristotle's prohibition of *metabasis* and the theory of subalternation of the sciences, see Steven J. Livesey, "*Metabasis:* The Interrelationship of the Sciences in Antiquity and the Middle Ages" (Unpublished PhD dissertation, The University of California, Los Angeles 1982), esp. chapter 1, and Richard D. McKirahan, "Aristotle's Subordinate Sciences," *British Journal for the History of Science* 11(1978): 197–220.

10. Walter Roy Laird, "The *Scientiae Mediae* in Medieval Commentaries on Aristotle's *Posterior Analytics*" (Unpublished PhD dissertation, University of Toronto, Centre for Medieval Studies, 1983), chapter VII, esp. at pp. 205–207. Laird argues that certain features of Paul's account—notably the contraction of the subalternating science through the super-added condition—could not have been taken from Aegidius's commentary, and probably were taken from Grosseteste. I have found no explicit reference to Grosseteste in Paul's commentary, and Grosseteste's theory was rather widely known at Oxford in the fourteenth century. If Paul acquired this idea while he was in Oxford, it may have come from many possible channels.

11. For my information concerning his biography, I rely primarily on the only extensive treatment of Antonius de Carlenis, G. Meersseman, "Antonius de Carlenis O.P., Erzbischof von Amalfi," *Archivum fratrum Praedicatorum* 3(1933): 81–131, supplemented where appropriate by more recent materials.

12. *I Rotuli dei lettori legisti e artisti dello studio Bolognese dal 1384 al 1799*, ed. Umberto Dallari, volume 1 (Bologna: Fratelli Merlani 1888), 10; volume 4 (1924), 31.

13. Ambrosius de Altamura, *Bibliothecae Dominicanae . . .* (Rome: Nicolai Tinassii 1677), 184; Vincentio Maria Fontana, *Sacrum theatrum dominicanum* (Rome: Nicolai Angeli Tinassii 1665), 51; Meersseman, pp. 103–105.

14. *Acta Capitulorum Generalium Ordinis Praedicatorum* III, ed. B. M. Reichert [Monumenta Ordinis Fratrum Praedicatorum Historica VII] (Rome: Typographia Polyglotta 1900), 244. He is also among the religious of the convent listed on 23 November 1439; see Celestino Piana, "La facoltà teologica dell' Università di Bologna nel 1444–1458," *Archivum Franciscanum Historicum* 53 (1960): 361–441 at p. 387 n. 2 and Piana, *Nuove ricerche su le Università di Bologna e di Parma.* (Quaracchi: Collegium S. Bonaventurae 1966), 317 n. 2.

15. Meersseman suggests (pp. 108–109) that the relationship between Antonius's theological training, his attendance at the Council, and the text of his commentary on the *Sentences* is not entirely conclusive. He speculates that Antonius attended as a colleague of Johannes de Monte Nigro, and that he lectured on the *Sentences* at Bologna during the year 1439–1440 (p. 103, n. 63). Thomas Kaeppeli [*Scriptores ordinis Praedicatorum Medii Aevi* vol. 1 (Rome: S. Sabina 1970), 109] offers the more inclusive dates 1439–1440.

16. Tommaso Kaeppeli, "Dalle pergamene di S. Domenico di Napoli," *Archivum Fratrum Praedicatorum* 32 (1962): 285–326 at pp. 302–303 (no. 650, 2 September 1447), 316 (2 September 1447), and in the following year he is once again referred to as 'magister in sacra pagina' (p. 303, no. 572).

17. The account, together with transcriptions of the relevant documents, is discussed in Meersseman, pp. 88–93.

18. See the excellent analysis by Denys Hay, *The Church in Italy in the Fifteenth Century* (Cambridge: Cambridge University Press 1977), esp. chapter 2 and appendix I.

19. *Registra Vaticana* 388, fol. 12$^{r-v}$, quoted at length in Meersseman, 89.

20. Jacobus Quétif and Jacobus Echard, *Scriptores Ordinis Praedicatorum Recensiti . . .* volume 1 (Paris: Ballard-Simart 1719), 820A.

21. *Questiones in libros I–II Analyticorum Posteriorum Aristotelis* I, q. 22; *infra*, 49. Concerning the circulation of these two texts, at least in one manuscript, see below, "Observations on the Manuscripts," xli–xlii.

22. The medieval collections of San Domenico, Bologna are discussed in P. V. Alce and P. A. D'Amato, *La Biblioteca di S. Domenico in Bologna* (Firenze: Leo S. Olschki 1966), esp. pp. 80–103, and M.-H. Laurent, *Fabio Vigili et les bibliothèques de Bologne au début du XVI$^e$ siècle* (Vatican: Biblioteca Apostolica Vaticana 1943): pp. 203–235 for the catalogue of books in the library before 1386, and pp. 11–107 for the inventory made by Fabio Vigili at the beginning of the sixteenth century. If one includes the libraries of the Friars Minor and the Augustinians, there was a substantial body of Antonius's sources available in Bologna during the fifteenth century. The medieval collection of San Domenico Maggiore, Naples, is more difficult to reconstruct. The earliest extensive catalogue is that published by Kaeppeli, "Antiche biblioteche Dominicane in Italia,"

*Archivum fratrum Praedicatorum* 36 (1955): 5–80 at pp. 29–53, most of which dates from the eighteenth century. Brito and Forlivio are not quoted in the two questions on subalternation, but are cited elsewhere in the commentary [Chicago, Newberry Library, Case MS 97,5, fol. 38$^{rb}$ (Brito) and fol. 6$^{va}$ and 23$^{ra-b}$ (Forlivio)]. In addition, Antonius cites Johannes de Janduno, whose views he refers to as 'ridiculus' (fol. 4$^{vb}$) and Alexander de Alexandria (fol. 38$^{rb-ra}$ and 41$^{rb}$). All of this serves to suggest that Antonius was using an unknown resource for his sources, perhaps his own educational experience earlier at Bologna. This may be the source for his knowledge of Paul of Venice, and perhaps his quotations of Girard of Bologna's commentary on the *Sentences* in his own *Sentence* commentary.

23. As has been mentioned in note 22, if one looks beyond the two edited questions in the *Questiones in libros Analyticorum Posteriorum*, one can add Jacobus de Forlivio, whose work dates from the beginning of the fifteenth century. But like Paul of Venice, he probably reflects a regional influence, since he was from Forlì and practiced medicine in Padua. In addition, at fol. 13$^{va}$, Antonius says, "Ad tertium, quod ista propositio 'animal est homo' reducitur ad primum modum secundum magistrum Paulum, sed quidam doctor Parisiensis arguit contra, quia omnis propositio per se est de omni, sed ista non est de (fol. 13$^{vb}$) omni. . .," but it is unclear whether this *doctor Parisiensis* was a contemporary of Paul or an earlier source whose position happened to speak to Paul's argument.

24. See Meersseman, 98.

25. The fourth conclusion, however, usually follows immediately from the others or restates an obvious aspect of the issue. For this reason, in the analysis that follows, I have usually omitted it.

26. See Meersseman, 99–103; G. Meersseman, "Een Vlaamsch Wijsgeer: Dominicus van Vlaanderen," *Thomistisch Tijdschrift* 1 (1930): 385–400, 590–592; Ulrich Schikowski, "Dominicus de Flandria O.P. († 1479), sein Leben, seine Schriften, seine Bedeutung," *Archivum Fratrum Praedicatorum* 10 (1940): 169–221; Franciszka Tuszyńska, "Materiały do Stanu Badań nad Filozofią Scholastyczną XV Wieku," *Studia mediewistyczne* 2 (1961): 5–99 at 20–21. I should like to thank my colleague, Professor Anthony S. Lis, for his assistance in translating this last source.

27. Tuszyńska *op. cit.*, 21, n. 118.

28. Prologue, q. 1; edn. pp. 5–7.

29. Prologue, q. 1; edn. pp. 7–10.

30. Prologue, q. 1; edn. p. 10.

31. Prologue, q. 1; edn. pp. 10–11.

32. Prologue, q. 1; edn. p. 11.

33. Prologue, q. 1; edn. pp. 11–13. The fourth conclusion (p. 13) is merely recapitulative.

34. Prologue, q. 2; edn. pp. 21–22.

35. Prologue, q. 2; edn. pp. 22–23. Antonius also cites Girardus de Bononia as a concurring authority to this opinion.

36. Prologue, q. 2; edn. p. 23. Antonius suggests that Girardus was responding directly to Peter Aureol, but on several counts, this seems to be impossible. Although the *Summa* was apparently written very late in his life [see Bartholomaeus F. M. Xiberta, "De Summa Theologiae magistri Gerardi Bononiensis ex Ordine Carmelitarum," *Analecta ordinis Carmelitarum* ann. XIV, v. 5(1923): 3–54 at 15–16], Peter Aureol's *Scriptum* was available only after 1317 [see *Scriptum super I Sententiarum*, ed. Eligius M. Buytaert, volume 1 (St. Bonaventure, NY: Franciscan Institute 1952–1956), xx and Katherine Tachau, *Vision and Certitude in the Age of Ockham. Optics, Epistemology and the Foundations of Semantics 1250–1345* (Leiden: E. J. Brill 1988), 88–89 n. 11]. More probably, Antonius conflated Aureol's arguments and Girard's discussions of the role of experience in subalternation (*Summa* q. 2, a. 3; Vatican, Borgh. lat. 27, fol. 7$^{vb}$, 8$^{va}$) as useful in his own discussion. Concerning the extended repercussions of this particular argument, see below, pp. xvii–xviii.

37. Durand de St. Pourçain, *In Petri Lombardi Sententias Theologicas Commentariorum libri IIII*, Prologue, q. 7, § 4; (Venice: Typographia Gerraea 1571; reprt. Ridgewood, NJ: Gregg Press 1964), fol. 12$^{ra}$.

38. Prologue, q. 2; edn. p. 23. In the same section, Antonius goes on to discuss the permanence of our theology and the residual status of faith in heaven, a point which divided Hervé and Durand and the other Scotists as well.

39. Prologue, q. 2; edn. p. 24.

40. Prologue, q. 2; edn. p. 25.

41. Prologue, q. 2; edn. p. 25.

42. On occasion, Antonius did break from this mold: in his *responsio* to the seventh principal argument in question 2, he argues against a position which he attributes to Landulfus Caracciolo; namely, that the subalternate science always has a composite subject. Such a criterion suggests that the subject of our theology, *Deus sub ratione deitatis*, should be composed of two things *per accidens*. But Antonius denies this criterion, at least in some cases: many commentators, he says, subalternate physics to metaphysics alone, and the example that Landulfus had given—perspective, which is subalternated simultaneously to physics and mathematics—is a peculiar one, since Thomas himself had argued that it was more natural than mathematical. There are two observations that might be made about this discussion. First, Antonius immediately buttresses his own discussion with remarks from Aegidius and Hervé. But beyond this, he takes no notice here of the fact that Thomas seems to have made contradictory remarks about the status of perspective, one in the commentary on the *Physics* (which Antonius cites), the other in his commentary on Boethius's *De Trinitate* (which he does not). This once again suggests a somewhat superficial discussion of the master's works in this question. However, one ought to note that Antonius does raise this problem in his commentary on the *Posterior Analytics* (I, q. 22; edn. p. 47) and attempts to reconcile the two positions.

43. Paul Oskar Kristeller, "Thomism and the Italian Thought of the Renaissance," in *Medieval Aspects of Renaissance Learning*, ed. and transl. Edward P. Mahoney (Durham, N.C.: Duke University Press 1974), 29–91 at p. 32.

44. Gregory of Rimini, *Lectura super primum et secundum Sententiarum*, Prologue, q. 1, ed. A. Damasus Trapp and Venicio Marcolino, t. 1 (Berlin-New York: De Gruyter 1981), 49.

45. Gregory of Rimini, *Lectura* Prologue, q. 1; *ed. cit.* p. 52. Concerning Gregory's discussion of theology as a science and this argument in particular, see Onorato Grassi, "La questione della teologia come scienza in Gregorio da Rimini," *Rivista di filosofia neo-scolastica* 58 (1976): 610–644, esp. at 631.

46. See above, p. xv.

47. Peter Aureol, *Scriptum super I Sententiarum*, Prooemium, sect. 1, A, 1, §5; B, 1a, §§23, 32; C, 1, §§134-135; ed. Eligius M. Buytaert, volume 1 (St. Bonaventure, NY: Franciscan Institute 1952–1956), 133, 139, 141, 172.

48. Johannes Capreoli, *Defensiones theologiae divi Thomae Aquinatis*, Prologue, q. 1, a. 2, § 2; ed. C. Paban and T. Pègues t. 1 (Tours: Alfred Cattier 1900; reprt. Frankfurt/Main: Minerva 1967), 6b–8a.

49. *Commentaria in IV libros Sententiarum*, quaest. praev. q. 1–4, quaesiuncularum solutio brevis, q. 1–7; in *Opera omnia* vol. 19 (Tournai: Typis Cartusiae S. M. de Pratis 1902), 58–83, 88–93. Concerning Denis's discussion of theology as a science, see Kent Emery, Jr., "Theology as a Science: The Teaching of Denys of Ryckel (Dionysius Cartusiensis, 1402–1471)," *Knowlege and Science in Medieval Philosophy. The Proceedings of the Eighth International Congress of Medieval Philosophy (SIEPM), Helsinki, 24–29 August 1987*, ed. R. Työrinoja, A. I. Lehtinen, D. Føllesdal, (Helsinki: Société Philosophique de Finlande 1990), Volume 3, pp. 376–388.

50. Thomas Claxton, *Scriptum super I Sententiarum*, Prologue, q. 2. The text survives in two of the three extant manuscripts of Claxton's commentary: Cambridge, Gonville and Caius College, MS 370(592), fols. 10$^v$-37$^r$, and Florence, BN Centrale, B.VI.340.1, fol. 6$^{vb}$-24$^{va}$. Concerning Claxton's discussion of subalternation, see my "Science and Theology in the Fourteenth Century: the Subalternate Sciences in Oxford Commentaries on the *Sentences*," *Synthèse* 83 (1990): 273–292 at 280–283.

51. For a more detailed discussion of Aristotle's theory of subalternation and its relationship to his prohibition of *metabasis*, see my *Theology and Science in the Fourteenth Century. Three Questions on the Unity and Subalternation of the Sciences from John of Reading's Commentary on the Sentences* (Leiden: E. J. Brill 1989), chapter 2.

52. *Questiones in libros Analyticorum Posteriorum Aristotelis* I, q. 17; edn. pp. 35–36.

53. *Questiones* I, q. 17; edn. p. 37.

54. *Questiones* I, q. 17; edn. p. 37.

55. *Questiones* I, q. 17; edn. p. 37.

56. *Questiones* I, q. 17; edn. p. 38.

57. *Questiones* I, q. 17; edn. pp. 38–39.

58. *Questiones* I, q. 22; edn. p. 46. In contrast to Aquinas's text [*In libros Posteriorum Analyticorum Expositio* I, lect. 25, no. 208; ed. R. Spiazzi (Turin: Marietti 1964), p. 231a], Antonius's formulation is "una scientia est dupliciter *subalternata*" (my italics). Although the abbreviation in the manuscript is quite clear, the fact that there is only one manuscript of the commentary might lead one to suggest that it was a scribal rather than conceptual error, but Antonius's remarks elsewhere make this defense somewhat weak.

59. *Questiones* I, q. 22; edn. p. 47.

60. Once again, one might suggest that Antonius has read his sources too hastily. He cites Walter Burley and Aegidius in this context, but both are quite clear in distinguishing pure mathematical disciplines from those subalternated to them. Aegidius [*In libros de physico auditu Aristotelis Commentaria* L. II, l. 3; (Venice: Bonetus Locatellus 1502), fol. 29$^{ra}$] reads: ". . . ut per mathematica intelligamus mathematica pure, cuiusmodi sunt geometria et arismetrica, et tunc quod dicitur non habet calumniam, quia cum perspectiva et astrologia sint medie inter geometriam et physicam cum medium magis approprinquet extremo quam extremum constat astrologiam et perspectivam magis esse physica quam geometria. . . ." A misreading of *arismetrica* for *astrologia* in this passage may have produced Antonius's rather curious position.

61. *Questiones* I, q. 22; edn. pp. 47–48.

62. *Questiones* I, q. 22; edn. p. 48.

63. *Questiones* I, q. 22; edn. pp. 48–49.

64. See Laird (note 10 above).

65. *Questiones* I, q. 22; edn. pp. 49–50.

66. *Quaestiones perutiles in D. Thomae Aquinatis Commentaria super libris Posteriorum Analyticorum Aristotelis.* (Venice 1587), p. 118b.

67. *Quaestiones perutiles*, p. 119b: "Considerandum est ulterius, quod genus potest accipi dupliciter: uno modo pro genere naturali, alio modo pro genere logicali, et hoc dupliciter. Uno modo pro genere praedicabili, alio modo pro genere subiecto sive pro genere scibili cui potest demonstrari inesse aliqua passio propria. Isto autem modo accipitur hic genus, et non duobus primis modis."

68. Quoted by Meersseman (1933) p. 100. In the same place, Meersseman has located and quoted the source from Antonius's commentary on the *Metaphysics*, which also has ties both to his commentary on the *Sentences* and the passage cited here in the *Posterior Analytics*.

69. In the second question to this *lectio*, Dominicus refers generically to the *schola Thomistarum*, but the reference is so imprecise that one cannot determine whether he considered Antonius within this category, and beyond this, the context suggests that he was appealing for support from this group.

70. *Quaestiones perutiles*, 159b.

71. *Quaestiones perutiles*, 160b–161a.

72. *Quaestiones perutiles*, 161b–162a.

73. *Quaestiones perutiles*, 162b–163a. In his *solutiones ad argumenta principalia*, Dominicus observes that some, like Albert of Saxony, have interpreted this as following from the aggregate character of scientific disciplines, something which he says has long been disputed in metaphysics (p. 164a).

74. *Quaestiones perutiles*, 163a. Dominicus recounts a four-fold distinction between the subalternating and subalternate sciences: (1) the way of knowing, (2) the thing known, (3) the more material or sensible emphasis of the subalternate science, and (4) the more formal and abstract emphasis of the subalternating science.

75. *Quaestiones perutiles*, 163a-b. Albert's three conditions for subalternation, properly speaking, are given as (1) that the subject of the subalternating science is contracted to that of the subalternate science; (2) that this contraction occurs through a material or extraneous differentia; and (3) that such contractions are not made *per passiones:* the subalternate subject is not merely a species of the subalternating subject. To these, Dominicus adds a fourth condition, that the subalternate science must take up its principles as proven in the subalternating science.

76. *Quaestiones perutiles*, 163b–164a. According to Dominicus, Albert of Saxony prescribed four types of subalternation: (1) in which the subject of one is part of the subject of the other, just as line is a part of magnitude; (2) in which one of the sciences operates on an uncontracted subject, while the other takes up that subject as it has been contracted essentially; (3) in which this contraction has been accomplished accidentally; (4) in which the conclusions of one science are demonstrated through the principles and conclusions of the other.

77. In addition to the issue of subject genus discussed above, one might add a reference to the similarity of *quia* and *propter quid* demonstrations *secundum analogiam sive proportionem* [*Quaestiones perutiles,* 162b], which recalls Antonius's remark in his *Questiones in libros Analyticorum Posteriorum* I, q. 17 [edn. p. 38].

78. *Questiones* I, q. 22 ad 11; edn. p. 51. Concerning this technique, see my "William of Ockham, the Subalternate Sciences, and Aristotle's Prohibition of *metabasis*," *British Journal for the History of Science* 18 (1985): 127–145 and "The Oxford Calculatores, Quantification of Qualities, and Aristotle's Prohibition of *metabasis*," *Vivarium* 24 (1986): 50–69.

79. Charles Schmitt, *Aristotle in the Renaissance* (Cambridge, MA: Harvard University Press 1983), esp. chapter 4.

80. Perhaps the richest example of this tendency remains Dionysius the Carthusian, whose works referred to earlier scholastics so extensively that he has become a thesaurus of source materials among modern historians. See Kent Emery, Jr., "Theology as a Science: The Teaching of Denys of Ryckel (Dionysius Cartusiensis, 1402–1471),"

*Knowlege and Science in Medieval Philosophy. The Proceedings of the Eighth International Congress of Medieval Philosophy (SIEPM), Helsinki, 24–29 August 1987*, ed. R. Työrinoja, A. I. Lehtinen, D. Føllesdal (Helsinki: Société Philosophique de Finlande, 1990), Volume 3, pp. 376–388.

81. Schmitt, pp. 90–91.

# OBSERVATIONS ON THE MANUSCRIPTS

## Questiones in IV libros Sententiarum.

This treatise is extant, so far as we know, only in the four manuscripts listed below:

1. **M** = Milano, Biblioteca Trivulziana 1682 (460), fol. $1^r$-$117^v$.

The hand is gothic, written in the second half of the fifteenth century. Paper; double columns of 46 lines; no initials or decoration. Of the four manuscripts, **M** possesses the greatest amount of marginal material, occasionally correcting, for the most part annotating the text. Although the watermarks suggest that the papers comprising the volume were Italian, the hand betrays some northern features, most notably in the superscripted i in *qui*. Originally owned by the Belgioioso family, the manuscript was passed to the Trivulzio collection probably in the nineteenth century, when some fifty volumes were brought as a dowry for Guilia Belgioioso upon her marriage to Gian Giacomo Trivulzio. The codex is described by Caterina Santoro, *I codici medioevali della Biblioteca Trivulziana* (Milano: Biblioteca Trivulziana, 1965) p. 307.

2. **R** = Rome, Biblioteca Casanatense 1025, fol. $1^{ra}$-$131^{rb}$.

Written in an Italian cursive hand, again from the second half of the fifteenth century; paper, bearing watermarks of an Italian corona design[1]; double columns of 42 lines; initial in gold with polychrome arabesque design (fol. $1^{ra}$), thereafter parted initials for Books II, III and IV; paragraph markers in red and blue. This is perhaps the most interesting of the four surviving copies, in part because it is the only manuscript to preserve the name of the scribe. On fol. $131^{rb}$ is written, "Expliciunt questiones super primo *Sententiarum* compilate a reverendissimo domino Domno A archiepiscopo, scripte per me Antonius de bagnolo . . . ," where a later hand has canceled 'compilate' and written above it 'edite'. On the basis of this colophon, and perhaps other information as well, P. T. Schiara, the eighteenth-century librarian of the Casanatense, suggested that this copy was produced while

the author was still alive,[2] and Meersseman concurred, arguing
that a different choice of words might have been used had Carlenis not been living. Meersseman also placed special emphasis
on the similarity between the colophon in B. Casanatense 1025
and the attribution made in the copy of Carlenis's *Questiones super
XII libros Metaphysicorum* preserved in Naples, BN VIII.G.75, written in 1456 by Egidius de Herct de Bruxella.[3] He suggested that
Antonius de Bagnolo, like Egidius, was a cleric in the service of
the Archbishop, whose duties included secretarial work.[4]

3. **O** = Oxford, Bodleian Library, Canon. misc. 573, fol. 172$^{ra}$-
377$^{ra}$.

This is the second part of a composite volume, written in an
Italian hand of the second half of the fifteenth century; paper;
double columns of 34–37 lines. The paper is Italian, bearing
among other things the watermarks of Italian licorne. It is, however, defective at the beginning, for it opens in the middle of the
*primum notandum* of question 1. The lost text is of some interest,
for while **O** reports a somewhat reliable version of the text subsequently, the portion of the *primum notandum* that remains is
rather different from that preserved in the other three
manuscripts.[5] Judging from the amount of text written in subsequent columns, it would appear that the preceding text would fit
into three of **O**'s columns, and allowing for its somewhat ex-
panded treatment of the *primum notandum*, certainly no more
than four columns have been lost. Yet the first quire of **O** is intact,
so that either the initial portion of the text was begun at the end
of a previous quire, or **O**'s exemplar was defective as well. The
latter is a distinct possibility, for **O** transmits the poorest text of
the four witnesses.

Still another aspect of **O** deserves to be noticed. In Meersseman's discussion of **R**, attention centered on the formula of attribution found in **R** and the copy of Carlenis's questions on the
*Metaphysics* preserved in Naples, BN VIII.G.75. Although
Meersseman did not mention it,[6] it is worth noting that the colophon of **O** repeats much the same formula: "Explicit liber super
quatuor libros Sententiarum compositus a reverendissimo domino Domno A. archiepiscopo Amalphitano, sacre theologie magistro egregio, ordinis Praedicatorum dignissimo, etc.. Amen." It
suggests that **R** and exemplar of **O** may have been related, and it
presents the possibility that **O** derives from a southern Italian location as well.[7] When we turn to the affiliation of the manuscripts, this issue will constitute a significant piece of evidence.

The manuscript was described by H. O. Coxe, *Catalogi codicum manuscriptorum Bibl. Bodleianae*, Pars III (Oxford: Clarendon Press, 1854); a more complete description is provided in Appendix 1.

4. **V** = Vatican, B. Reginensis lat. 392, fol. 105$^r$-224$^v$.

Written on paper in an Italian hand of the second half of the fifteenth century; double columns of 47–53 lines. It is defective at the end, halting abruptly in book III.

Of the four surviving copies, the provenance of **V** is the most certain and straightforward. By accident, the upper margin of fol. 105$^r$ preserves the number "353" which marks it as having been in the Vatican library between 1508 and 1513, when Fabio Vigili de Spolète made an inventory of the collection (now Vat. lat. 7135–7136). It may also have been included among the earlier inventories of the late fifteenth century, although this cannot be determined with any precision.[8]

• • • • •

On the basis of a collation of the manuscripts, the transmission of the text can be summarized as follows:

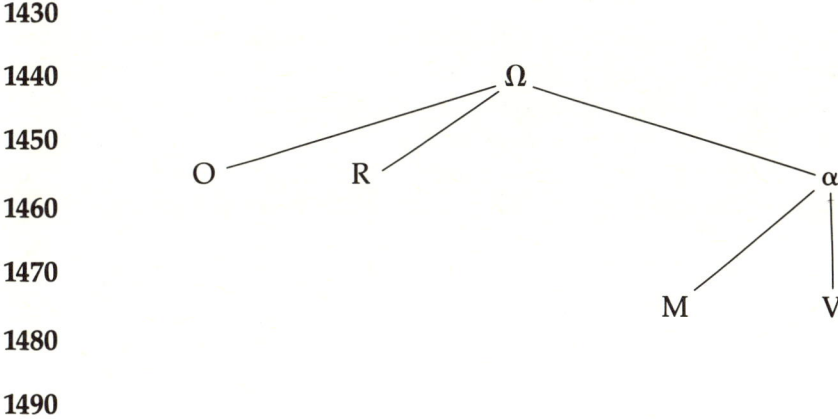

**1430**

**1440**

**1450**

**1460**

**1470**

**1480**

**1490**

The manuscripts of the *Questiones in IV libros Sententiarum* can be classified in three branches which descend from a single archetype. If Meersseman's argument concerning **R** is correct, it constitutes the oldest of the extant manuscripts, perhaps from the 1450s. Likewise, the watermarks in **O** would make an early date for it possible as well.[9] Meersseman's suggestion that **R** was copied under the potential supervision of Antonius himself leaves the possibility that some changes were made in the text. This, combined with the variation in the *primum notandum* of question 1 in **O**, may suggest that **O** was an earlier version that was replaced by **R**, and that the remaining manuscripts reflect this revision as well. But having said this, one should also note that in other respects **O** and **R** are very similar, that this small portion of text is slim evidence on which to base such a hypothesis, and more significantly that it is possible the formula in the colophons of **O**, **R**, and Naples BN VIII.G.75 may be similar because they were taken over from a still earlier exemplar, and hence do not necessitate an early date for these copies. In view of this indeterminate situation, we have suggested that all copies derive from a single version, and that in accord with Meersseman's suggestion (which as we will see receives some support from the colophon in the sole surviving manuscript of the *Questiones in libros I–II Analyticorum Posteriorum Aristotelis*), **O** and **R** were produced before Carlenis died.

O and **R** read together against the other two manuscripts on over one hundred occasions in these two questions from the prologue. But more significantly, they read against each other on almost the same number of occasions. None of the other three could have been copied from **O**, since it omits significant readings present in the others [eg. in Q. 2, p. 22, lines 24–25]. Nor could **O** have been copied from any of the other three: its frequent divergences from **M** and **V** (both in omissions and more significantly additions) are apparent from the apparatus, and it reads correctly with **M** and **V** against **R** on several occasions as well (Q. 1, p. 9, line 13, subalternate **MOV** ] subalternante R; Q. 1, p. 15, line 11: erit **MOV** ] dicit R; Q. 2, p. 27, line 6: duobus **MOV** ] rationibus R). And as has been noted above, **O** preserves a significantly different *primum notandum* in Q. 1, in which **M, R,** and **V** quote a different passage from Hervé Natalis than does **O**.

The conclusion that must be drawn is that **O, R,** and **M** and **V** represent different witnesses to the archetype. The existence of α is inferred from the frequent agreement of **M** and **V** against **O** and R; in the two questions from the prologue, **V**, for example, reads with **M** and against **O** and **R** over 150 times, while it reads with **O** and **R** and against **M** fewer than 20 times. In fact, it diverges from **M, O,** and **R** an even greater number of times, twenty nine. Yet neither **M** nor **V** was copied from the other. In Q. 1, p. 12, lines 4–5, **V** omits distinguitur a fidei, que non est habitus discursivus, which **M** contains correctly against **O** and **R** as well, following the quotation in Hervé Natalis.[10] Conversely, **V** reads correctly with **O** and **R** at Q. 2, p. 22, line 19 de eo de against **M**'s de eodem; and shortly thereafter, **V** reads with **O** theologus against **M** and **R** theologicus (Q. 2, p. 25, line 1).

The picture that emerges from this examination is that while **O** and **R** may be older witnesses and may have been produced (at least in the case of R) in the proximity of the author (if Meersseman is correct), **M** and **V** cannot be discounted as significant witnesses of the archetype. This is apparent particularly in passages where Antonius quotes from sources: on several occasions, **M** and **V** preserve the reading found in the source.[11]

## Questiones in libros I–II Analyticorum Posteriorum Aristotelis

This treatise is extant apparently in a single manuscript: **C** = Chicago, Newberry Library, Case MS 97,5, fol. 2^{ra}-47^{vb}.

Written in humanistic cursiva tending to currens by Petrus de Afelatro in Naples, 1468; paper; double columns of 46 lines; initial fol. 1^{ra} decorated, otherwise initials in red and blue. The codex also contains Thomas Aquinas's commentary on the *Perihermenias* in the same hand. In the eighteenth century, the codex was in the possession of Hyacynthus a Murano, a Dominican friar of Calabria; it was acquired by Bernard Rosenthal from Hoepli[12] and subsequently sold to the Newberry Library in 1962. It has been described by Paul Saenger, *A Catalogue of the Pre-1500 Western Manuscript Books at the Newberry Library* (Chicago: University of Chicago Press, 1989), 188–189.

Petrus de Afelatro, a professor of philosophy and medicine at the University of Naples, is reported to have written a commentary on Aristotle's *Metaphysics*.[13] On fol. 47^{va}, the colophon reads, "Expliciunt questiones super primum et secundum *Posteriorum* edite per reverendissimum dominum Antonium quondam archiepiscopum Amalfitanum scripte per me Petrum anno domini millesimo cccc° lxviii mensis augusti, complete in nocte sancti Martini. Scripsit has questiones Petrus de Afelatro de Neapoli 5° anno sui studii in hiis artibus." Two things ought to be emphasized about this colophon. First, it constitutes the only tangible evidence that Antonius's work received attention in schools outside the Order, although the existence of other manuscripts of his works and the evidence that still other copies existed at one time would suggest that there was some limited, perhaps regional, influence. Second, and more significantly, is the similarity of the formula to those employed in manuscripts of the commentaries on the *Sentences* and the *Metaphysics*. As we have suggested above, it seems that very early, the manuscript tradition of Antonius's works established this formula, but Petrus has added "quondam" to his colophon, because of course by 1468 Antonius had been dead for eight years. Its omission in the other three manuscripts may give some, albeit limited, strength to the suggestion that **O** and **R** were produced before 1460.

Finally, although so far as we know, the *Questiones in libros I–II Analyticorum Posteriorum Aristotelis* exists in only this manuscript, evidence from **M** may suggest that at least one other copy circulated in the fifteenth century. In the margin, folio $3^{va}$, the scribe has written

Sed posset etiam responderi, quia dato quod subiectum alicuius scientie sit compositum ex duobus, sed quia ex illis aggregatis resultatur quidam conceptus simplex, non oportet ipsam scientiam duabus subalternari, et hoc etiam docet iste doctor infra, questionibus libri *Posteriorum*, quare etc.

The context here in the response to the seventh principal argument is Antonius's restatement of Landulfus Caracciolo's argument that subalternate sciences are subordinated to two superior sciences simultaneously. The source for the theory is Landulf's commentary on the *Sentences*, and indeed we have no evidence that Landulf wrote a commentary on the *Posterior Analytics*. The doctor referred to in the marginal note must be Antonius himself, leading us to conclude that **M** was followed by the commentary on the *Analytics*. Although it is possible that the copy was **C** itself, one should note that the hands are entirely different, that the page and written space vary somewhat, and that **M** continues with another anonymous text in the same hand that produced the commentary on the *Sentences*.

### Editorial Procedures

The critical apparatus is organized traditionally.[14] I have attempted to provide all the significant variations presented in the five manuscripts. In addition, wherever practical, I have provided marginal notations, both of corrections to the text and scribal annotations. The chief exception to this occurs in the *Questiones in libros I–II Analyticorum Posteriorum Aristotelis*, where the scribe has inserted divisions of the text that would be superfluous in this transcription. Unfortunately, the more substantive marginalia in this manuscript are occasionally obscured by the binding; I have indicated these portions in the apparatus / . . . /.

Wherever possible, I have attempted to indicate direct quotations by double quotes. However, I have used some discretion in comparing Antonius's text with that of the source, since in most cases the sources have not been edited critically, and in the case of Hervé Natalis, the edition is only a partial one; where appropriate, these quotations are designated by single quotes.

## Notes

1. For information concerning the watermark design, I should like to thank Dott. Marta Corsanego, who reports that the design on the first leaf of each bifolium is similar, but not identical to that of Briquet no. 4705. However, even if one considers the related figures 4714 and 4717, most of these papers are associated with a northern Italian provenance of an earlier period.

2. On the flyleaf, Schiara has written ". . . Scriptum fuit per Antonium de Bagnolo, Auctore vivente. . . ."

3. G. Meersseman, "Antonius de Carlenis O.P., Erzbischof von Amalfi," *Archivum fratrum Praedicatorum* 3(1933): 81–131, at 94–95, 106–107. Until the new catalogue of the Casanatense manuscripts is complete, Meersseman's description of MS 1025 is for the most part serviceable.

Because of the connection between the Casanatense manuscript, the Naples copy of the commentary on the *Metaphysics,* and the Oxford copy of the commentary on the *Sentences* (see below), I reproduce the incipit in Naples, BN VIII.G.75, fol. 1$^r$: "Incipiunt questiones super libros metaphysice, a reverendissimo domino A. archiepiscopo amalphitano sacre theologie magistro egregio" (Meersseman, p. 94).

4. Meersseman, p. 107.

5. For a transcription of this fragment, see Appendix 2.

6. In his 1933 article, Meersseman was apparently unaware of the Oxford manuscript; by 1935, Pelzer had brought it to his attention, and in a brief supplementary article, he announced the incipit and colophon without additional analysis. See "Ergaenzung zum Schrifttum des Antonius de Carlenis von Neapel O.P.," *Archivum Fratrum Praedicatorum* 5 (1935): 357–363 at 360.

7. I have not been able to inspect the Naples copy of the commentary on the *Metaphysics,* something which might shed further light on this connection. It seems that the commentaries on the *Metaphysics* and the *Sentences* may have circulated together; the library of S. Domenico Maggiore in Naples possessed a copy of each, and in 1765, the librarian, Vincenzo Gregorio Lavazzoli reported "Abbiamo di esso li Commentari in 4 libros sententiarum et super Metaph. Aristotelis"; see T. Kaeppeli, "Antiche biblioteche Domenicane in Italia," *Archivum fratrum Praedica-*

*torum* 36(1966): 5–80 at pp. 37, no. 111 and 52 no. 19. This raises the further possibility that two of the extant commentaries on these works were held together, and perhaps were produced together. Unfortunately, Meersseman did not report information about watermarks of the papers in the Naples manuscript, and other codicological evidence that could settle this question is not readily available.

8. Concerning the inventories made by Fabio, see Jeanne Bignami Odier, *La Bibliothèque Vaticane de Sixte IV à Pie XI* [Studi e testi 272] (Vatican: B. Apost. Vaticana 1973), 27–28; M. Bertolà, *I due primi Registri di prestito della Bibl. Apost. Vaticana* (Vatican: B. Apostolica Vaticana 1942), pp. xii–xiii; and M. H. Laurent, *Fabio Vigili et les bibliothèques de Bologne au début du XVI*ᵉ *siècle d'après le ms. Barb. lat. 3185* [Studi e testi 105] (Vatican: B. Apostolica Vaticana 1943), xviii, xx. Concerning the 1475 and subsequent inventories (Vat. lat. 3964 and others), see Bertolà, *I due primi Registri*.

9. The licorne designs recorded by Briquet similar to that found in O range in date from 1426 to 1436; the té design is similar to Briquet tracings from 1454–1456. See Briquet vol. 2, nn. 9957, 9960, 9971, and 14089.

10. In Q. 2, p. 27, line 10 also, M alone reads correctly mathematicis against methaphysicis in O, R, and V.

11. This is especially true of Antonius's favorite source, Hervé Natalis.

12. Based on a personal conversation with Mr. Rosenthal.

13. Concerning Petrus de Afelatro, see Francesco Torraca *et al.*, *Storia del Università di Napoli* (Naples: Riccardo Ricciardi Editore 1924), 185, 325, 330, 336. Gianguiseppe Origlia [*Istoria dello Studio di Napoli* (Naples: Giovanni di Simone 1753), 246–247] reported that among the faculty of philosophy and medicine was "Pier d'Afeltro, che ci lasciò i Commentarj sulla Metafisica d'Aristotele. . . ."

14. With modifications for differences in subject matter, I have followed many of the suggestions made by Stephan Kuttner, "Notes on the Presentation of Text and Apparatus in Editing Works of the Decretists and Decretalists," *Traditio* 15 (1959): 452–464. When necessary, this has been supplemented by A. Dondaine, "Abbréviations latines et signes recommandés pour l'apparat critique des éditions de textes médiévaux," *Bulletin de la Société internationale pour l'étude de la philosophie médiévale* 2 (1960): 142–149 and "Variantes de l'apparat critique dans les éditions de textes latins médiévaux," *Bulletin de la Société internationale pour l'étude de la philosophie médiévale* 4 (1962): 82–100.

# ANTONIUS DE CARLENIS DE NEAPOLI
## *QUESTIONES IN IV LIBROS SENTENTIARUM*
## PROLOGUS, QQ. 1 & 2

# Conspectus Siglorum

M = Milano, Bibl. Trivulziana 1682 (XV)
O = Oxford, Bodleian Library, Canon. Misc. 573 (XV) (mutil.)
R = Roma, Bibl. Casanatense 1025 (XV)
V = Vaticana, Reginsis lat. 392 (XV)

• • • • •

| | |
|---|---|
| *add.* | addidit |
| *canc.* | cancellavit |
| *conj.* | conjectura |
| *corr.* | correxit |
| *del.* | delevit |
| *expand.* | expandit |
| *ins.* | inseruit |
| *lin.* | linea(m) |
| *marg.* | margine |
| *dext.* | dextro |
| *sinist.* | sinistro |
| *obscur.* | obscuratus |
| *om.* | omisit |
| *ras.* | rasura |
| *suppl.* | supplevit |
| *trans.* | transposuit |
| < > | omissiones supplevimus |

# \<QUESTIO 1:
## 'UTRUM DOCTRINA SACRA SIT SCIENTIA'>*

Circa prohemium primi[†] Sententiarum, primo[‡] queritur utrum doctrina sacra sit scientia. Probatur quod non, quia omnis scien-
5  tia procedit ex principiis per se notis. Sed theologia procedit ex articulis fidei, qui non sunt per se noti. Ergo etc.[§]

Secundo, non potest esse maior evidentia in conclusione quam in principiis. Sed notitia principiorum theologie non est nisi ex fide, que non est nisi ex non evidentibus. Assumptum probatur,
10  quia[#] non est maior evidentia in effectum quam in causa.

Tertio, "propter quod[**] <est> unumquodque et illud[††] magis," I *Posteriorum*,[1] et ad hoc proprie inducitur ab Aristotele. Si ergo // (V fol. 105rb) deducta ex principiis creditis sciuntur, ergo de principiis erit simul[‡‡] fides et scientia, quod est implicare
15  contradictionem.

Quarto, capio habitum theologicum in mente prophetarum et sanctorum; manifestum est quod ibi fuit facta revelatio, et de eodem et secundum idem est scientia[§§] nostre theologie[##] cum scientia illorum. Sed apud illos per revelationem, erat sola fides.
20  Ergo de theologia nostra non est alius habitus nisi fidei.

Quinto, habitus cuius oppositum[***] est heresis, ille est fidei. Sed habitus contradicens illis que[†††] habentur in theologia est[‡‡‡] heresis. Ergo habitus theologie est fidei, et non scientie.

Sexto, scientia proprie dicta resolvit[§§§] usque[###] ad principia
25  per se nota scienti; sed theologia non resolvit usque ad principia[****] per se nota habentibus theologiam.

Septimo, si dicatur scientia ex articulis fidei tanquam ex principiis, quanto magis aliquid polleret fide que// (V fol.

---

*M, R fol. 1ra; V fol. 105ra.
[†]primi *om.* V.
[‡]primo *om.* R.
[§]Ergo etc. *om.* R.
[#]quia M, V ] quod R.
[**]propter quod M ] quod R ] *om.* quod V.
[††]et illud R ] tale et illud M, V.
[‡‡]erit simul *transp.* R.
[§§]scientia R, V ] scire M.
[##]nostre theologie *transp.* R.
[***]oppositum M, V ] obiectum R.
[†††]que M, V ] qui R.
[‡‡‡]fidei *add.* V.
[§§§]resolvit R ] resolvitur M, V.
[###]usque *om.* R.
[****]principia *om.* R.

4

# <QUESTION 1: 'WHETHER THEOLOGY IS A SCIENCE'>QUESTION 1

Concerning the prologue of the first book of the *Sentences*, it is asked first whether theology is a science. It is proved that it is not, since every science proceeds from principles known per se. But theology proceeds from the articles of faith, which are not known per se. Therefore.

Second, the evidence in the conclusion cannot be greater than [the evidence] in the principles. But knowledge of the principles of theology is only by faith, which can only be from non-evidents. The assumption is proved, since the evidence is not greater in the effect than in the cause.

Third, "that on account of which something is so must be even more so itself," according to the first book of the *Posterior Analytics*, and this is properly used inductively by Aristotle. If therefore things deduced from believed principles are known, both faith and knowledge will be from principles at the same time, which is to imply a contradiction.

Fourth, I take the theological habit to be in the mind of the prophets and the saints; it is clear that at one time there was a revelation, and from it and following it there is the science of our theology and that of the blessed. But among the prophets, through the revelation, there was only faith. Therefore, in our theology, there is only a habit of faith .

Fifth, the habit opposed to heresy is faith. But the habit opposed to those things which are held in theology is heresy. Therefore the theological habit is of faith, not of science.

Sixth, science properly speaking reduces to principles that are known per se to the knower. But theology does not reduce to principles known per se to those with the habit of theology.

Seventh, if it is said that the science is from the articles of faith as from principles, the more something avails itself of faith for the

105va) est articulorum, tanto magis polleret theologia, quod est contra Augustinum XIV* *De Trinitate* c. <. . . . >,[†] ubi dicit quod plurimi pollent fide sed non scientia.[2]

Octavo, de contingentibus non est scientia. Sed in theologia
5  tractatur quod Deus creavit, vel quia creat,[‡] et sic[§] de aliis[#] contingentibus.

Nono, scientia // (R fol. 1rb) est de aliquo quod habet causam. Sed theologia est de Deo, qui[**] non habet causam.

Decimo, theologia est de singularibus, ut de Christo et aliis, et
10  sic[††] non est scientia.

Undecimo, quia theologia etiam[‡‡] non videtur[§§] esse[##] <aliud> nisi notitia consequentialis. Ergo non est scientia. Probatur antecedens, quia theologia non videtur esse[***] aliud nisi notitia deductiva ex auctoribus aliquarum conclusionum.

15  Duodecimo, scientia et fides non possunt esse de[†††] eodem. Sed iste habitus non potest esse[‡‡‡] nisi fidei. Ergo nullo modo potest dici scientia.


In contrarium, est quia // (M fol. 1rb) dicitur sapientia et scien-
20  tia, et patet ex doctrina sancti doctoris, I[a] pars, q. 1, articulo 2 et 6[§§§].[3]


### <Quattuor notanda>


Primo, notandum est[###] quod in hac questione facit difficul-
tatem, in[****] hoc, quod dicit sanctus doctor in libro preallegato,
25  quia dicit eam esse scientiam et sapientiam, sed Scotus[4] et Petrus Aureoli[5] magis eam dicunt debere dici sapientiam quam scientiam, ut infra patebit.[6] Sed Herveus, in questionibus quas facit[††††] de scientia theologie ad declarationem dictorum sancti doctoris, in secundo articulo[7] expresse tenet

*XIV ] IV R, M, V.
[†]< . . . > *lacuna* R, M, V.
[‡]quia creat R ] creet M, V.
[§]sic *add.* V.
[#]aliis *om.* M, V.
**qui M, V ] que R.
[††]et sic M, R ] ergo V.
[‡‡]etiam *om.* M, V.
[§§]videtur R ] videretur M, V.
[##]esse *om.* R.
***esse M, V ] etiam R.
[†††]de M, R ] in V.
[‡‡‡]esse R ] dici M ] *om.* V.
[§§§]et 6 *om.* R, V.
[###]est *om.* M, V.
****in *add.* V.
[††††]facit R ] fecit M, V.

articles, the more theology should avail itself; this is contrary to what Augustine says in *On the Trinity*, Book XIV, <part 1, ch. 3>, where he says that most people avail themselves of faith, not of science.

Eighth, there is no science of contingents. But that God has created or that he is creating and other such contingents are discussed in theology.

Ninth, science concerns that which has a cause. But theology is about God, who has no cause.

Tenth, theology concerns singulars, as for example Christ and others, and so it is not science.

Eleventh, because theology does not seem to be anything more than knowledge of consequences. Therefore, it is not a science. Proof of the antecedent: because theology seems to be nothing more than the deductive knowledge of some conclusions from authors.

Twelfth, there cannot be science and faith of the same thing. But this habit cannot exist except from faith. Therefore, it cannot be called science in any way.

To the contrary, it is called Wisdom and science, and it is clear from the teaching of St. Thomas, Iᵃ part, question 1, articles 2 and 6.

<Four Notations>

First, note that in this question there is a difficulty concerning what St. Thomas says in the aforesaid book, since he says that it is science and wisdom, but Scotus and Peter Aureoli say that it should be called wisdom rather than science, as will be shown below. But in the question he formulated to declare the teachings of St. Thomas about the science of theology, Hervé holds expressly in the second article

quod ex premissis, quarum una est credita et altera* scita, non
potest inferri nisi credita, quia sicut se habent necessarium et
contingens in inferendo necessarium, ita evidens et inevidens ad
inferendum evidentiam. Sed ex contingenti et necessario, non se-
5    quitur nisi conclusio contingens et non necessaria per Philoso-
phum I *Posteriorum*.[8] Et idem probatur ex II *Posteriorum*,[9] quia ad
probandum predicatum de subiecto in eo⁺ quod quid est, oportet
accipere ambas quod quid est. Similiter‡ etiam in I *Posteriorum*,[10]
quod oportet scire propositiones per se notas. Et§ idem Herveus
10   // (V fol. 105vb) in 3 articulo[11] dicit, quod ad scientiam subalternam non sufficit arguere ex tantum creditis. Et attende quod
multi concedentes# quod possit salvari theologiam esse scientiam per hoc, quod est subalterna, credunt se sufficienter sal-
vasse. Unde dicit Herveus**[12] quod sanctus doctor de hoc sensit⁺⁺
15   in‡‡ hac declaratione, quod principia subalterne§§ possunt dupliciter accipi.## Uno modo, quantum ad quia est; alio modo,
quantum ad propter quid est. Et*** loquor de scientiis subalternis
humanitus adinventis modo dico quod principia scientie subalterne, quantum ad quia est, non sunt tantum credita etiam in
20   illa⁺⁺⁺ scientia subalterna,‡‡‡ ymmo sunt per se // (R fol. 1va) nota,
ut non accipiat ea a scientia subalternante quantum ad quia est,
quantum autem ad propter quid est; principia scientie subalternate sunt tantum credita in scientia subalternata precise accepta,
et sunt demonstrata in scientia subalternante. Dat ibi exemplum
25   de§§§ perspectivo respectu geometrie; postea subdit Herveus.[13]
Nunc dico quod scientia subalternata,### quantum ad hoc quod
est proprie scientia, procedit ex principiis per se notis, ita ut sint
per se nota quantum ad quia est, et sic facit evidentiam de suis

---

*altera *om.* M, V.
⁺eo *om.* M, V.
‡similiter R ] sequitur M, V.
§et *om.* R.
#quod multi concedentes R ] quia concedentes M, V.
**dicit Herveus *transp.* R.
⁺⁺sensit M, R ] sentit V.
‡‡in M, V ] cum R.
§§subalterne ] subalternante R ] subalterna M, V.
##accipi M, V ] capi R.
***et *om.* R.
⁺⁺⁺illa M, V ] ista R.
‡‡‡subalterna M, V ] subalternata R.
§§§sp *add. sed canc.* M.
###subalternata R ] subalterna M, V.

that from premises one of which is believed and the other known, nothing can be inferred unless it is believed, because just as the necessary and the contingent are related in inferring the necessary, so the evident and the inevident [are related] in the inference of the evident. But from the contingent and the necessary nothing follows but the contingent, not the necessary, as the Philosopher [shows in] *Posterior Analytics*, Book I. The same is proved from the second book of the *Posterior Analytics*, since to prove the predicate of the subject in its essential nature, one ought to grasp both in their essential nature. The same is found in the first book of the *Posterior Analytics*, because one must know such propositions on their own terms. And Hervé says the same in article 3, that in the subalternate science it is not sufficient to argue only from matters that are merely believed. And note that many who concede that [the proposition], "theology is a science," can be saved on the ground that it is a subalternate science believe that it has been saved, but barely. Whence Hervé says that St. Thomas felt the same way in this declaration, that subalternate principles can be taken in two ways, insofar as they state the fact, and insofar as they state the reasoned fact. And I speak of subalternate sciences created by humans in the way I say that principles of the subalternating science, insofar as they state the fact, are not simply believed in the subalternating science but are known in themselves, in such a way that one should not take them to be stating the fact in the subalternating science, but rather the reasoned fact; principles of the subalternate science are believed [and] adopted in the subalternate science, and demonstrated in the subalternating science. In that context, he gives the example of perspective with respect to geometry, and later Hervé does so as well.

Now I say that a subalternate science, insofar as it is properly scientific, proceeds from principles that are known in themselves, so that they are known in themselves as the fact, and in this way the science produces evidence of its

conclusionibus; non autem procedit ex eis inquantum sunt credita quantum ad propter quid est,* quia ut sic non facerent evidentiam de suis conclusionibus, sed credulitatem, nec per
consequens est scientia proprie dicta. Et hec dicta Hervei; vide in
5 questione sequenti.[†14]

Secundo, notandum quod Scotus,[15] Petrus Aureoli[16] et
Franciscus[‡] super I *Sententiarum*[17] reprobant[§] hoc quod[#] dicit
sanctus doctor,[18] quod articuli fidei sint principia huius scientie.
Et Scotus[19] et Franciscus[20] reprobant hoc quod dicit,[**21] quod est
10 scientia[††] subalternata scientie[‡‡] beatorum. Et[§§] // (M fol. 1va)
Scotus distinguit[##] quod tripliciter potest sumi // (O fol. 172rb)
theologia,[22] sed Franciscus de Mayronis clarius distinguit[***]
quod theologia quattuor modis potest accipi[23]: Uno modo, pro
habitu quem habuerunt illi, quibus revelata est. Secundo modo,
15 pro illo habitu qui generatur in nobis ex studio // (V fol. 106ra)
sacre[†††] scripture. Tertio modo, pro illo habitu qui deducitur ex
veritatibus contentis[‡‡‡] in sacra scriptura. Quarto modo, pro habitu declarativo et defensivo fidei. Et de[§§§] primo modo,[###] di-
mittit sub dubio quis fuerit ille habitus; secundo

---

*est *om.* R.

[†]Et hec dicta Hervei; vide in questione sequenti. *om.* R; O, which is defective at the beginning, commences with a variant of the *primum notandum* (fol. 172ra); see Appendix 2.

[‡]Scotus, Petrus Aureoli et Franciscus V ] Petrus Aureoli O, R ] Scotus, Aureoli Petrus et
Franciscus M.

[§]reprobant M, V ] reprobat O, R.

[#]quod *add. in marg.* O.

[**]quod dicit M, V ] dicit R ] dictum O.

[††]est scientia *transp.* O.

[‡‡]scientie *om.* O, R.

[§§]Et O, M, V ] Ut R.

[##]distinguit O, R, V ] dicit *corr. in marg.* distinxit M.

[***]distinguit M, O, V ] dicit R; M *Marg.*: "<pro>hemio ar. 14."

[†††]sacre M, O, V ] sacro R; theologie *add. sed canc.* O.

[‡‡‡]et *add.* O, R.

[§§§]de *om.* R.

[###]modo *add.* M, V.

conclusions. It does not, however, proceed from principles believed as the reasoned fact, since in such a circumstance they
would not produce evidence of its conclusions, but [rather] belief,
nor consequently would there be science properly speaking.
These quotations are from Hervé; note them in the next question.

Second, note that Scotus, Peter Aureoli, and Francis in the first
book of the *Sentences* offer a counterargument to what St. Thomas
says, that articles of faith are principles in this science. And
Scotus and Francis argue against what he says, that it is a science
subalternated to the science of the blessed. And Scotus makes the
distinction that theology can be taken in three ways, but Francis
of Mayronnes makes the clearer distinction that it can be taken in
four senses: in one way, for the habit which they hold, by which
it has been revealed. Second, for the habit which is produced in
us by the study of sacred scripture. Third, for that habit which is
deduced from the truths contained in sacred scripture. Fourth,
for the habit that declares and defends the faith. He leaves as-
doubtful the habit that arose in the first way; of the second

modo, dicit quod est fides; tertio modo, dicit quod non est scientia, quia aut deducitur ex utraque credita, et tunc conclusio erit
tantum credita, aut ex una credita et alia per se nota, et tunc eodem modo. Si autem accipiatur quarto modo, sic est una pars no-
5   bilissima et summa methaphysice, et tandem concludit quod
theologia vel accipitur ut in se, vel quoad nos, in se capitur pro
habitu illo quem potest facere obiectum in* intellectum propor-
tionatum. Secundo modo, capitur pro illo quem intelligens potest
percipere. Primo modo, est scientia, ut patet. Secundo modo,
10  non, quia non† possumus habere certam notitiam de obiecto, ut
sciamus certam veritatem scientie theologice ut theologice sunt.
Et idem dicit Landulfus[24] hic, qui dicit quod Thomas dicit quod
theologia est scientia proprie, quia accipit principia credita que
ipsis beatis sunt scita, et arguit contra hoc sicut supra dictum est.
15  Et ibidem‡ ponit[25] dictum sancti doctoris, quod ponit in IIª IIᵉ, q.
1, articulo 5,§[26] ubi dicit // (O fol. 172va) quod non potest fides et
scientia# de eodem esse,** per Apostolum *Ad Hebreos* 11.[27] Et ad
idem // (R fol. 1vb) ponit opinionem Godofrido,†† *Quolibet* VIII, q.
7‡‡, quod fides contradicit scientie.[28] Et contra hoc arguit
20  Landulfus,[29] quia§§ actus demonstrandi, cum non includat de-
meritum, non adnichilat fidem, que## est habitus infusus.

Item, apostoli viderunt Christum crucifixum, qui*** est unus
articulus fidei, et tamen habuerunt fidem, et Paulus, pro parvo
tempore quo vidit††† Deum per essentiam, non credo quod per-
25  didit fidem.

Item, notitia matutina et vespertina in eodem intellectu

---

*in *om.* O.
†non *om.* R.
‡ibidem M, O, V ] idem R.
§q. 1, articulo 5 R ] articulo prima, q. 1 O ] q. 1, articulo < . . . > M, V.
#fides et scientia O, R ] scientia et fides M, V.
**esse *add.* M.
††M *marg.:* 'Godo.'
‡‡7 ] 27 M, O, R, V.
§§quia M, O, V ] quod R.
##que R ] qui M, O, V.
***qui M, V ] quia O, R.
†††vidit R, V ] videt M, O.

way, he says that it is faith; of the third way, he says that it is not science, since either it is deduced from [premises] both of which are believed, and then the conclusion will only be believed, or [it is deduced] from [premises] one of which is believed and the other known in itself, and then the same thing applies. If however, it is taken in the fourth way, it is the most noble and highest part of metaphysics, and at length he concludes that whether theology is taken as it is in itself, or insofar as we are concerned, it refers to that habit which can make the object proportionate to the intellect. In a second way, it is taken as that which the one knowing can perceive. According to the first way, it is science, as is clear. According to the second way, it is not, since we cannot have certain knowledge of the object in such a way that we would know a certain truth of theological science precisely as theological. Landulf says the same thing here, when he says that Thomas says that theology is properly science, since he accepts as believed principles that are known to the blessed, and argues against this, as has been said above. In the same place, he lays down the point which St. Thomas makes in IIᵃ IIᵉ, q. 1, article 5, where he says that there cannot be faith and science of the same object, according to the Apostle in *Hebrews* chapter 11. And to the same issue, he presents the opinion in Godfrey's *Quodlibet* VIII, question 7, that faith is the opposite of science. And Landulf argues against this, since the act of demonstrating, when it is not against charity, does not destroy faith, which is an infused habit.

Again, the apostles saw Christ crucified, which is one of the articles of faith, and nevertheless they had faith, and I do not believe that Paul, during the short time in which he saw God in his essence, lost faith.

Again, knowledge of the morning and the evening are in the

beati sunt. Sed quia ista non faciunt ad propositum, dimitto.
Manifestum est autem ex dictis, quod Franciscus[30] cum aliis
Scotis dicunt quod iste* habitus est fidei, // (V fol. 106rb) cum quibus concordant alii,[†] dicentes quod est opinio Petri Aureoli,[31]
5   quod est habitus declarativus. Sed Herveus, ut in primo notando
dictum est, dicit declarando sancti doctoris,[32] quod est scientia
largo modo dicta, et cum expositione eius concordat Ioannes de
Neapoli, dicens[33] quod sanctus doctor, in questione disputata 14,
articulo 9 in responsione ad tertium,[34] dicit 'quod scientia subal-
10  ternata non est proprie scientia, nisi in quantum // (M fol. 1vb)
continuatur in[‡] eodem subiecto cum scientia subalternante.' Et
idem Ioannes reprobat[35] illud quod aliqui dicunt, quod[§] ex hac
positione sancti doctoris, 'quod de ratione scientie subalternate[#]
est quod procedat ex principiis non in ipsa notis, sed in aliqua
15  superiori, scilicet in** subalternante.' Et sic concludit quantum ad
notitiam, quod quia[††] necesse est quod habeatur // (O fol. 172vb)
in subalternata,[‡‡] et quod hec doctrina non est scientia proprie
accepta, et sic 'potest dici scientia largo modo, sicut scientia
iuristarum.'[36] Et articulus Parisiensis dicit[§§] quod 'error est dicere
20  quod nichil penitus sciatur propter scire theologiam.'[37] Et concedit idem Ioannes,[38] quod capiendo prima documenta[##] theo-
logica secundum se, de eis est*** scientia,[†††] licet nobis non sint

---

*iste M, O, V ] ille R.
†concordant alii *transp.* M.
‡in O, R, M ] cum V.
§quod *add.* M, V.
#subalternate M, O, V ] subalternante R.
**in *om.* R.
††quod quia *transp.* O, R.
‡‡subalternata M, R, V ] alternata O.
§§dicit *add.* R, V.
##documenta M, R, V ] doctrina O.
***est *om.* O.
†††*ante* licet: licet non *add. sed canc.* R.

same beatified intellect. But since this is not pertinent, I leave it aside. However, it is clear from what has been said, that Francis, together with other Scotists, say that this habit is one of faith, with which others agree, saying that this is the opinion of Peter Aureoli, that it is a declarative habit. But Hervé, as has been said in the first notation, says in asserting [the position] of St. Thomas, that it is science in a broad sense, and John of Naples is in agreement with his position, when he says that St. Thomas in his disputed question 14, article 9, at the response to the third principal argument says "the subalternate science is not properly science except insofar as it is continuous with the subject of the subalternating science." And John argues against what some say about this view of St. Thomas, "that the nature of a subalternate science is that it proceed from principles that are not known in the science itself, but in some superior science, namely, the subalternating science." And this is what he concludes about such knowledge, that since it must be held in the subalternate [science], and this [sacred] doctrine is not science in its proper sense, "it can be called science in a broad sense, just like legal science." And an article of Paris says that "It is an error to say that nothing is known completely by knowing theology." And John concedes the same, that in grasping primary theological doctrines in themselves, there is science of them, although they are not

evidentia. Si vero loquamur quantum ad suas conclusiones, sic
est scientia proprie* et stricte. Scimus enim quod conclusiones
huius doctrine necessario sequuntur ex suis principiis, et sic
potest dici scientia quinque modis: primo, quantum ad ea que
5    habet[†] communia cum methaphysica; secundo, secundum se,
licet[‡] non quoad nos; tertio, quantum[§] ad notitiam possibilitatis
suorum principiorum; quarto, largo modo, quia multa scimus de
articulis fidei que non scit[#] simplex fidelis; quinto, potest dici scientia consequentiarum, sicut dicit theologus quod conclusiones
10   theologie** necessario sequuntur ex suis principiis.

Tertio, notandum quod apud doctores hic communiter
ponitur[††] opinio Henrici,[‡‡39] qui posuit aliquod lumen medium
inter lumen[§§] glorie // (R fol. 2ra) et lumen fidei, et quantum ad
hoc quod sit possibile, concordat Scotus[40] quod potest communi-
15   cari de deitate notitia abstractiva, et quod illa non sit beatifica. Et
de hoc vide clare[##] Franciscum in*** I *Sententiarum*, in // (V fol.
106va) prohemio q. 15.[41] Sed sanctus doctor de hoc non inquirit,
et sic pertranseundum est, maxime cum[†††] loquimur de habitu
acquisito qui communiter acquiritur, et non de possibili tantum.
20   Inde[‡‡‡] quia secundum istos etiam in via, potest communicari
notitia propter quid articulorum fidei, et[§§§] quod habitus
theologici[###] // (O fol.

*scientia proprie *transp.* M, V.
[†]habent *corr.* M.
[‡]*ante* non: quod O.
[§]quantum M, O, V ] quo R.
[#]scit M, V ] sit O, R.
**theologie M, O, V ] theologice R.
[††]ponitur M, O, V ] ponit R.
[‡‡]Henrici *corr. in marg.* M.
[§§]medium inter lumen *add. in marg.* O.
[##]clare *om.* M, V.
***in *om.* O, R.
[†††]cum *om.* M, R, V.
[‡‡‡]Inde O, R ] vide M, V.
[§§§]et M, V ] sed O, R.
[###]theologici M, V ] theologice O, R.

evident to us. If indeed we speak of its conclusions, there is science properly and strictly speaking. For we know that the conclusions of this doctrine follow necessarily from its own principles, and so it can be called science in five ways: first, insofar as it has something in common with metaphysics; second, in itself, although not to us; third, with respect to the knowledge of the possibility of its own principles; fourth, in a broad sense, since we know many things about the articles of faith that the simple believer does not know; [and] fifth, it can be called a science of consequences, just as the theologian says that the conclusions of theology follow necessarily from its own principles.

Third, note that among the doctors, Henry [of Ghent's] opinion is commonly applied, that there is some light intermediate between the light of glory and the light of faith, and to the extent that this is possible, Scotus agrees that one can have an abstractive cognition of God, though this would not be a beatific cognition. And about this, see the clear reference in Francis's commentary on book I of the *Sentences*, in the prologue, question 15. But St. Thomas did not investigate this, and so we must pass over it, especially since we are speaking of an acquired habit that is commonly acquired, and not simply of one that is possible. Then since even for wayfarers there can be scientific knowledge of the articles of faith, theological habits

173ra) declarentur* ab habitu fidei tali et† fundarentur.‡ Sed necessarium est dicere quod§ habitus theologie non est solum adhesivus, sicut est fides, quia# paganus posset esse theologus** et
tamen non habere†† habitum infusum fidei. Sed Franciscus,[42] qui
5    dicit habitum theologicum esse fidem, non intelligit de fide infusa, sed de habitu fidei, quem etiam aliquis ex studio acquirit
vel‡‡ accipit.

Quarto, notandum quod§§ scientia non est consequentiarum
tantum, ut fuit opinio quorundam, ut recitat Herveus in prima
10   questione, articulo quarto,[43] ubi dicit distinguendo quod
scientia## duobus modis potest se habere ad consequentiam:
uno modo, ut habeat consequentiam // (M fol. 2ra) tanquam per
se obiectum, et sic spectat ad logicam; secundo modo, ut se
habeat*** ad consequentiam sicut ad modum cognoscendi, et iste
15   modus distinguendi Hervei est conformis modo exponentium
libros *Posteriorum*, et colligitur ex II *Methaphysice*,[44] qui distinguit
de logica docente et de logica††† utente.

### <Alique conclusiones>

Prima conclusio: habitus theologie non est fides. Probatur, quia
20   habitus scientificus non est fides. Sed‡‡‡ theologia est
aliquomodo§§§ habitus scientificus. Et confirmatur: fides est de
non evidentibus, et ideo sanctus doctor, I pars, q. 1, bene
concedit[45] quod certitudo est communis tam scientie quam fidei.
Tamen quantum ad istam### conclusionem, ego multum ad-
25   hereo dictis Hervei, qui materiam**** istam discussit in questione
2, articulo 2,[46] ubi concludit sic. 'Et mihi videtur quod theologia
nichil aliud est quam quidam habitus creditivus faciens fidem de
hiis, que formaliter et implicite in fide et articulis continentur
propter credulitatem previorum creditorum, qui sunt articuli // (V
30   fol. 106vb) fidei, in quo distinguitur fides a scientia, et cum

---

*declarentur M, V ] declaratur O, R.
†et M, O, V ] quia R.
‡fundarentur M, V ] fundaretur O, R.
§quod M, O, V ] quia R.
#quia M, V ] et O, R.
**theologus M, V ] theologicus O, R.
††habere M, O, R ] haberet V.
‡‡acquirit vel *om.* O, R.
§§ista *add.* R.
##scientia *om.* O.
***habeat O, R ] habet M, V.
†††de logica *om.* M, V.
‡‡‡sed *om.* O, R.
§§§necessario *add.* R.
###istam O, M, V ] illam R.
****materiam O, R ] manifeste M, V.

are made clear by such a habit of faith and founded upon it. But one must say that a habit of theology is not [a habit of] adherence alone, like faith, since a pagan can be a theologian and still not have the infused habit of faith. But Francis, who speaks of the theological habit as faith, does not mean infused faith, but the habit of faith that one acquires or accepts by study.

Fourth, note that it is not a science of consequences only, as was the opinion of some, as Hervé recalls in the first question, article 4, where he makes the distinction that science can relate to a consequence in two ways: in one way, as it has the consequence as its proper object, and so this refers to logic; [and] in the second way as it regards the consequence as a manner of knowing. And the latter way is Hervé's distinction in conformity with his way of interpreting the *Posterior Analytics*, as is gathered from the second book of the *Metaphysics*, which distinguishes between logic as a doctrine [*logica docens*] and logic in use [*logica utens*].

<Some Conclusions>

First conclusion: the habit of theology is not faith. Proof: a scientific habit is not faith, and theology is in some way a scientific habit. Confirmation: faith is of things that are not evident, and for that reason St. Thomas concedes in the first part, question 1, that certitude is common to both science and faith. Nevertheless, insofar as this conclusion is concerned, I adopt for the most part what Hervé has said in question 2, article 2, where he concludes in this way: "It seems to me that theology is nothing other than a certain believing habit that produces faith about things that are contained formally and implicitly in the faith and its articles, and does so on the basis of the belief of previous believers. These are the articles of faith, in which faith is distinguished from science

eis convenit, quia in hoc, quod est habitus // (O fol. 173rb) creditivus, convenit cum fide et distinguitur a scientia, que est habitus evidens. In hoc autem quod est habitus discursivus,
distinguitur a fide, que non est habitus // (R fol. 2rb)
5 discursivus,'* et sic in eodem articulo, distinguit†⁴⁷ quia‡ fides
potest accipi dupliciter: uno modo pro habitu qualitercumque
creditur alicui, sine auctoritate divine absolute, et in se, sive ut
deducto ex alia auctoritate, et isto modo concludit theologiam
esse§ fidem proprie dictam. Alio modo, potest accipi fides magis
10 stricte pro habitu quo assentimus alicui revelato absolute, non ut
deducto ex alio, et theologia pro habitu faciente assentire alicui
credito deducto ex tali auctoritate, et sic theologia et fides# dicunt
habitus distinctos realiter, sicut habitus principiorum et habitus
conclusionum deductarum ex principiis illius scientie. Et hec
15 sunt verba Hervei.

Secunda conclusio: scientia theologie dicit habitum discursivum. Probatur, quia non dicit tantum habitum adhesivum, et nec
habitum simplicem innativum, et** in hoc concordat Petrus Aureoli, qui dicit quod ista†† scientia dicit alium habitum a‡‡ fide,
20 quia frustra esset§§ quod## in ea studetur,*** et quod iste††† habitus non est adhesivus tantum, et quod est habitus declarativus,
et hoc est verum non solum de‡‡‡ theologia et§§§ secundum se, de
qua Scotus et omnes sequentes eum### dicunt quod est vera
scientia; sed etiam**** de theologia nostra, de qua etiam as-
25 serimus quod est scientia, licet non secundum omnem sui
condicionem.††††

Tertia conclusio: patet quomodo sanctus doctor intelligit

---

*In hoc autem quod est habitus discursivus, distinguitur a fide, que non est habitus *om.*
O; In hoc autem quod est habitus discursivus, et convenit cum scientia, que est habitus //
[R fol. 2rb] discursivus R; distinguitur a fidei, que non est habitus discursivus *om.* V.
†distinguit O, R, V ] distinxisset *corr. in textu* M.
‡quia M, O, R ] quod V.
§esse R ] et M, O, V.
#et fides M, O, V ] fidei R.
**nec habitum simplicem innativum et *add. in marg.* O.
††ista O, M, V ] illa R.
‡‡a O, R ] de M, V.
§§esset O, M, V ] est R.
##quod *add.* M, V.
***studetur M, O, V ] studitivus(?) R.
†††iste M, O, V ] ille R.
‡‡‡de M, R, V ] ut O.
§§§et *om.* R.
###sequentes eum *transp.* M ] eum omnes sequentium V.
****etiam *om.* R.
††††condicionem M, O, R ] conclusionem V.

and has elements in common with it. Insofar as it is a believing habit, [theology] has elements in common with faith and is distinguished from science, which is a habit based on evidence; insofar as it is a discursive habit, it is distinguished from faith, which is not a discursive habit." Thus in this same article, [Hervé] proposes the distinction that faith can be taken in two ways: in one way for the habit which is believed by someone in whatever way, without divine, absolute authority and either in itself or as it is deduced from another authority, and in this way he concludes that theology is faith properly speaking. In another way, faith may be taken more strictly as the habit by which we assent to something revealed absolutely, not as it is deduced from something else, and theology, as the habit inducing assent to a certain belief derived from such an authority. In this way, theology and faith are said to be habits that are really distinct, just as are the habits of principles and the habits of conclusions deduced from those principles in the science itself. So the words of Hervé.

The second conclusion: the science of theology refers to a discursive habit. Proof: since it does not refer only to a habit of assent [habitus adhesivus], nor is it a simple innate habit. And in this, Peter Aureoli agrees, when he says that this science refers to a habit different from faith, since [otherwise] one would study it in vain, and [when he says that] this habit is not one of assent only, and that it is a habit of clarification. And this is true not only of theology as it is in itself (about which Scotus and all those following him say that it is a true science), but also of theology as it is in us, which we also assert is a science, although not in all respects.

The third conclusion: It is clear in what way St. Thomas understands

quod* scientia theologie est habitus ac- // (M fol. 2rb) quisitus.
Probatur ex supradictis, quia non ponitur lumen infusum ut
ponit Henricus,[48] nec ponitur fides infusa, sed habitus deducti-
vus ex principiis ad conclusiones.

5      Quarta conclusio: patet quomodo sanctus doctor accipit quod
dicatur theologia et[†] scientia.

&lt;Ad argumenta principalia&gt;

Ad primum argumentum,[‡] quod illud argumentum // (O fol.
173va) concludit quod non dicatur proprie scientia, quia manifes-
10    tum est quod theologia[§] habet pro principiis articulos fidei.

Ad secundum, quod illud argumentum etiam[#] concludit quod
non est proprie scientia, ut patebit in sequenti questione.[49]

Ad tertium, quod** similiter probat quod non potest esse evidentia sufficiens in deductis ex principiis creditis, sed non probat
15    quin cum exercitio et disputationibus ad explicationem termino-
rum et articulorum et defensiones et probabilium rationum ad
miracula clarius veritas habeatur.

Ad quartum, quod necesse est dicere quod sit[††] aliquis alius habitus ab habitu qui est per revelationem puram,[‡‡] sicut fuit in pro-
20    phetis. // (R fol. 2va) Et hic est dubitatio, si notitia revelata et
acquisita de eodem obiecto sint[§§] unius speciei, quia Petrus Aureoli in secunda questione super prohemium dicit quod sic,[50] et
Franciscus de Mayronis dicit quod non,[51] sed sanctus doctor
ponit virtutes infusas et acquisitas distingui specie, quia sunt ad
25    alium et alium finem,[52] sed si essent ad eundem finem, ipse non
diceret quod revelatum et acquisitum distinguerentur[##]
specie,*** sicut nec oculus qui sit miraculose videns et qui est naturaliter videns.[†††]

Ad quintum, quod oppositum theologie est ignorantia, et non
30    heresis; et hoc est bonum argumentum contra illos qui dicunt
quod habitus theologie est fides, maxime si dicatur[‡‡‡] esse
fidem[§§§]

---

*patet quomodo sanctus doctor intelligit quod *om.* M, V; est *add.* O.
[†]et *om.* M, V.
[‡]*Marg.:* "Ad primum argumentum" M.
[§]theologia M, V ] scientia O, R.
[#]etiam *add.* M, V.
**quod *om.* M, V.
[††]sit *om.* M, V.
[‡‡]revelationem puram *transp.* M, V.
[§§]de *add.* O, M, V.
[##]distinguerentur M, O, V ] distingueretur R.
***specie *om.* M, V.
[†††]et qui est naturaliter videns *om.* R.
[‡‡‡]dicatur R ] dicant M, O, V.
[§§§]fidem *om.* M, V.

that the science of theology is an acquired habit. This is proved from the discussion above, since an infused light is not posited in the way that Henry [of Ghent] posits it, nor is an infused faith posited, but rather a habit that is deductive, [proceding] from principles to conclusions.

The fourth conclusion: the way in which St. Thomas speaks of theology as a science is obvious.

<Responses to the Principal Arguments>

To the first argument: the argument concludes that it is not science in a strict sense, since it is clear that theology takes the articles of faith as its principles.

To the second: the argument concludes that it is not science in a strict sense, as is clear in the following question.

To the third: [the argument] proves similarly that there cannot be sufficient evidence in things deduced from principles that are believed, but it does not prove that a truth is not held more clearly by an exercise or disputation to explain terms and articles and to defend the miraculous with probable reasons.

To the fourth: this is a habit different from one that exists purely through revelation, as there was in the prophets. And here there is a doubt whether revealed knowledge and acquired knowledge of the same object is of the same species, since Peter Aureoli says in the second question of his prologue that it is, and Francis of Mayronnes that it is not. But St. Thomas establishes that infused and acquired powers are distinct in species, since they have different ends. But if they had the same end, he would not say that the revealed and the acquired should be specifically distinct, just as the eye that sees something miraculous would not be distinct from the eye that sees something natural.

To the fifth: the opposite of theology is ignorance and not heresy. And this is a good argument against those who say that the habit of theology is faith, especially if it is said to be only in-

infusam tantum. Sed Scotus[53] et Franciscus[54] intelligunt quod sit quedam fides acquisita.

Ad sextum, quod secundum Herveum[55] possumus* dicere quod modo suo resolvit unam conclusionem in alteram premis-
5 sam, sicut dicendo quod filius // (O fol. 173vb) Dei verum habuit sanguinem quia veram habuit humanitatem, et quod fuit vere coloratus quia habuit verum corpus mixtum quia est vere† homo. Et hoc tandem probatur per argumentum, quia filius Dei homo factus est, et illa‡ talis facultas§ deducendi et declarandi non de-
10 bet dici consequentialis tantum, nec pure fides, sed habitus declarativus et defensivus,# ut supra dictum est.[56]

Ad septimum, quod ille habitus,** licet causetur ex principiis // (V fol. 107rb) fidei, tamen requiritur exercitium et actus frequentatus, qui causant istum habitum.

15 Ad octavum, quod†† principaliter ex necessariis, et si est ex‡‡ contingentibus, est ex illis§§ in ordine ad necessaria.

Ad nonum, quod omnes premisse possunt habere rationem cause,## ut patet II *Physicorum*[57] et V *Methaphysice*,[58] et etiam non inconvenit*** dicere quod in conceptibus dictis de Deo, unus di-
20 catur prior††† alio,‡‡‡ saltem§§§ secundum rationem.

Ad decimum, // (M fol. 2va) quod est### de singularibus sub quibusdam rationibus universalibus et regulis et de**** quibusdam singularibus gratia exempli.

---

*possumus M, R, V ] possemus O.
†vere O, R ] verus M, V.
‡illa O, R ] ista M, V.
§facultas R ] facilitas M, O, V.
#et defensivus *om.* O.
**habitus *om.* O.
††quod M, V ] qui O, R.
‡‡est ex V ] est de O, R ] ex M.
§§illis M, V ] tali O, R.
##cause *om.* O.
***inconvenit R ] convenit O, M, V.
†††prior *corr. in textu* M.
‡‡‡alio O, R ] altero M, V.
§§§saltem *om.* V.
###est *om.* M.
****de *om.* M, V.

fused faith. But Scotus and Francis mean that it is a kind of acquired faith.

To the sixth: according to Hervé we can say that in its own way the science reduces one conclusion into another premise, as when saying that the Son of God had true blood because he was truly human, and that he was truly colored because he had a real composite body because he was truly human. And finally, this is proved by the argument that the Son of God was made man, and that such a capacity for deducing and declaring must not be said to be consequential alone, nor purely faith, but a declarative and defensive habit, as has been said above.

To the seventh: although that habit arises from the principles of faith, nevertheless exercise and frequent activity are required to generate this habit.

To the eighth: [knowledge is] principally from necessaries, and if it is from contingents, it is from them as they are related to necessaries.

To the ninth: every premise can have the formality of a cause, as is clear from *Physics* II and *Metaphysics* V, nor is it improper to say that in concepts said of God, one is said to be before another, at least according to reason.

To the tenth: [theology] is of singulars considered under certain universals and rules, and of some singulars merely by way of example.

Ad undecimum, negatur quod sit notitia consequentialis, ut dictum vel* declaratum est in quarto notando.

Ad duodecimum, quod sanctus doctor, in II$^a$ II$^e$ q. 1, articulo 5,[59] dicit quod non potest esse fides et scientia de eodem, quia *Ad*
5  *Hebreos* 11,[60] fides est argumentum non apparentium, et idem arguit Godofridus,[61] *Quolibet* VIII,$^†$ q. 7,$^‡$ sed Durandus hoc improbat.[62] Et similiter Landulfus[63] et Giraldus[64] hic dicentes quod fides$^§$ // (R fol. 2vb) est inevidens negative, quia non facit evidentiam, non tamen facit$^#$ inevidentiam.** Et vide bene de hoc in se-
10  cundo // (O fol. 174ra) *Quolibet* Hermani de Maio,[65] 14 questione, et infra declarabitur[66] ubi erit$^{††}$ magis ad propositum. Sed pro nunc, dictum est quod iste habitus theologicus non est simpliciter fides, et tenendo$^{‡‡}$ viam predictorum, qui dicunt quod potest esse fides et scientia de eodem, facilius esset respondere. Et ean-
15  dem viam sicut sanctus doctor tenet[67] hic Giraldus,[68] qui ponit et reprobat expositionem illorum qui dicunt illud I *Ad Corinthianos* c. 13,[69] quod fides evacuabitur,$^{§§}$ esse verum non quantum ad habitum, sed quantum ad enigma, et concludit quod fides et scientia opponuntur ratione$^{##}$ obiectorum mediorum et actuum.

---

*dictum vel *om.* M, V.
$^†$VIII *om.* M, V.
$^‡$7 ] 27 M, O, R, V.
$^§$quod fides *redupl.* V.
$^#$tamen facit *transp.* R.
**non tamen facit inevidentiam *om.* M, V.
$^{††}$erit M, O, V ] dicit R.
$^{‡‡}$tenendo O, R ] tenent M ] tenere V.
$^{§§}$evacuabitur M, V ] evacuabatur O, R.
$^{##}$ratione *om.* O.

To the eleventh: I deny that it is knowledge of consequences, as has been said or declared in the fourth notation.

To the twelfth: St. Thomas says in II<sup>a</sup> II<sup>e</sup>, question 1, article 5 that there cannot be faith and knowledge of the same conclusion, since according to *Hebrews* 11, faith is the evidence of things unseen, and Godfrey maintains the same in *Quodlibet* VIII, question 7, but Durand argues against this. Similarly, Landulf and Girard say that faith is negatively nonevident, since it does not produce evidence, nor does it produce nonevidence. About this, see the second *Quodlibet* of Herman of Maio, question 14; this will be dealt with below, where it will be more pertinent. But for now, it has been said that this habit of theology is not faith absolutely, and, taking the path of those mentioned previously, who say that there can be faith and knowledge of the same conclusion, the response is easier [to understand]. And in this place, Girard holds the same position as St. Thomas, when he maintains and reproves those who cite I *Corinthians* chapter 13, that faith will pass away, [saying that this] is true, not as it concerns the habit [of faith], but as it concerns its obscurity, and he concludes that faith and knowledge are distinguished on the ground of their intermediate objects and acts.

## Notes

1. *Analytica Posteriora* I.2 72$^a$29; *Aristoteles Latinus* IV.1–4, ed. L. Minio-Paluello and B. G. Dod (Bruges-Paris: De Brouwer 1968), 9.

2. *De Trinitate* XIV.i.3; ed. W. J. Mountain, *Corpus Christianorum*, ser. lat. 50A (Turnholt: Brepols 1968), 424.

3. *Summa theologiae* I, q. 1, a. 2, 6; ed. P. Caramello (Turin: Marietti 1952), vol. 1, pp. 3, 5–6.

4. Scotus, *Ordinatio* Prol., pars 4, q. 1–2, §213; *Opera Omnia* vol. 1 (Vatican: Polyglottis 1950), 146.

5. Petrus Aureoli, *Scriptum super Primum Sententiarum*, Prooemium, sect. 1, a. 2d; ed. E. Buytaert, volume 1 (St. Bonaventure, NY: Franciscan Institute 1952–1956), 166ff.

6. *Infra*, 7–10.

7. Hervé Natalis, *Defensa doctrinae S. Thomae* prima pars, I, a. 2; ed. Engelbert Krebs, *Theologie und Wissenschaft nach der Lehre der Hochscholastik* (Münster i. W.: Aschendorffsche Verlagsbuchhandlung 1912), 7*.

8. *Analytica Posteriora* I.6 75$^a$12–17; *Aristoteles Latinus* IV.1–4, p. 201.

9. *Analytica Posteriora* II.8 93$^a$25–30; *Aristoteles Latinus* IV.1–4, p. 81.

10. *Analytica Posteriora* I.6 75$^a$30–31; *Aristoteles Latinus* IV.1–4, p. 19.

11. Hervé, *Defensa doctrinae* prima pars, I, a. 3; ed. Krebs, 9*.

12. Hervé, *Defensa doctrinae* prima pars, I, a. 3; ed. Krebs, 10*.

13. Hervé, *Defensa doctrinae* prima pars, I, a. 3; ed. Krebs, 10*.

14. Antonius de Carlenis, *Questiones* . . . Prol. q. 2; ed. p. 24.

15. Scotus, *Reportata Parisiensia* III, d. 24, q. unica; *Opera omnia* vol. 23 (Paris: Vivès 1894), 446–447.

16. Petrus Aureoli, *Scriptum* I, Prooem., sect. 1 B, art. 1a, § 24–29; ed. Buytaert, volume 1, pp. 141–142.

17. Franciscus de Mayronis, *In IV libros Sententiarum*, Prologue, q. 14; (Venice: O. Scotus 1520), fol. 8$^{vb}$P-Q.

18. *Summa theologiae* I, q. 1, a. 2; *ed. cit.* vol. 1, p. 3.

19. Scotus, *Reportata Parisiensia* III, d. 24, q. unica; *Opera Omnia* vol. 23, pp. 447–449.

20. Franciscus de Mayronis, *In IV libros Sententiarum*, Prologue, q. 14; *ed. cit.* fol. 8$^{vb}$P-Q.

21. *Summa theologiae* I, q. 1, a. 2; *ed. cit.* vol. 1, p. 3.

22. Scotus, *Ordinatio* Prologue, pars 3, q. 3, § 196–205; *Opera Omnia* vol. 1 (Vatican: Polyglottis 1950), 133–138.

23. Franciscus de Mayronis, *In IV libros Sententiarum* Prologue, q. 14 (Venice: O. Scotus 1520), fol. 8$^{va-b}$.

24. Landulfus Caracciolo, *Super I Sententiarum* Prologue, q. 2, a. 3; Vienna, Nationalbibliothek 1496, fol. 5$^{ra}$.

25. Landulfus, *Super I Sententiarum* Prologue, q. 4, a. 1; fol. 7$^{va}$.

26. *Summa theologiae* II, ii, q. 1, a. 5; *ed. cit.* vol. 2, p. 7b.

27. *Ad Hebreos* xi.1.

28. Godofridus de Fontibus, *Quodlibetum* VIII, q. 7; *Les Philosophes Belges. Textes et études 4. Le huitième quodlibet de Godefroid de Fontaines*, ed.

J. Hoffmans (Louvain: Institut supérieur de philosophie de l'Université 1924), 73.

29. Landulfus, Prologue, q. 4, a. 1; fol. 7$^{va}$.

30. Franciscus de Mayronis, *In IV libros Sententiarum* III, d. 23–33, q. unica, a. 6; (Venice: O. Scotus 1520), fol. 171$^{rb}$.

31. Petrus Aureoli, *Scriptum* I, Prooem. sect. 1 B, art. 3c, § 112; ed. Buytaert, volume 1, pp. 164–165.

32. Hervé, *Defensa doctrinae* prima pars, I, a. 3; ed. Krebs 10–11.

33. Joannes de Neapoli, *Questiones disputatae* q. 18, punct. 2 (Neapoli: C. Vitalis 1618; reprt. Ridgewood, NJ: Gregg Press 1966), 148a.

34. Aquinas, *De veritate* q. 14, a. 9 ad. 3; *Quaestiones disputatae* v. 1, ed. R. Spiazzi (Turin-Rome: Marietti 1964), p. 298a. The quoted passage is, however, from Joannes de Neapoli.

35. Joannes de Neapoli, *ed. cit.* p. 148a-b.

36. Joannes de Neapoli, *Quaestiones disputatae* q. 18, punct. 3; *ed. cit.* p. 154a.

37. *Chartularium Universitatis Parisiensis*, ed. H. Denifle and E. Chatelain, volume 1 (Paris: ex typis fratrum Delalain 1889), no. 473; p. 552, # 153.

38. Joannes de Neapoli, *Quaestiones disputatae* q. 18, punct. 4; *ed. cit.* p. 156a.

39. Henricus Gandavensis, *Summae quaestionum ordinariarum* art. 13, q. 6; (Paris: J. Badius 1520; reprt. St. Bonaventure, NY: Franciscan Institute 1953), fol. 94$^r$.

40. Scotus, *Reportata Parisiensia* III, d. 24, q. unica, ad 7; *ed. cit.* vol. 23, p. 459b.

41. Franciscus de Mayronis, *In IV libros Sententiarum* Prol., q. 15; *ed. cit.* fol. 9$^{rb}$-10$^{ra}$.

42. Franciscus de Mayronis, *In IV libros Sententiarum* III, d. 23–33, q. unica, a. 3; *ed. cit.* fol. 171$^{ra}$A.

43. Hervé, *Defensa doctrinae* prima pars, I, a. 4; ed. Krebs, 12*.

44. *Forsitan Metaphysica* II.3 995$^a$14–15; *Aristoteles Latinus* XXV.2, ed. G. Vuillemin-Diem (Leiden: Brill 1976), 39–40.

45. *Summa theologiae* I, q. 1, a. 5; *ed. cit.* vol. 1, pp. 4–5.

46. Hervé, *Defensa doctrinae* prima pars, I, 1., a. 9; ed. Krebs, 47*.

47. Hervé, *Defensa doctrinae* prima pars, I, 1., a. 9; ed. Krebs, 47*.

48. Henricus Gandavensis, *Summae quaestionum ordinariarum* art. 13, q. 2 (Paris: J. Badius 1520; reprt. St. Bonaventure, NY: Franciscan Institute 1953), fol. 91$^r$.

49. Antonius de Carlenis, *Questiones* . . . Prol. q. 2; ed. p. 24.

50. *Forsitan Scriptum* I, Prooem., sect. 1, B, art. 1e, § 65–74; ed. Buytaert, 151–153.

51. Franciscus de Mayronis, *In IV libros Sententiarum* III, d. 23–33, q. unica, a. 6; *ed. cit.* fol. 171$^{rb}$G.

52. *Summa theologiae* I, ii, q. 63, a. 4; *ed. cit.* vol. 1, p. 278b.

53. Scotus, *Quaestiones in librum tertium Sententiarum* d. 23, q. unica; *Opera omnia* vd. 15 (Paris: Vivès 1894), 7a–8b.

54. Franciscus de Mayronis, *In IV libros Sententiarum* III, d. 23–33, q. unica, a. 1; *ed. cit.* fol. 170$^{va-b}$.

55. Hervé, *Defensa doctrinae* prima pars, I, a. 2; ed. Krebs, 7*.

56. *Supra*, 12.

57. *Physica* II.3 195$^a$18–20; *Aristoteles Latinus* VII.1, fasc. 2, ed. Fernand Bossier and Jozef Brams (Leiden-New York: E. J. Brill 1990), 59.

58. *Metaphysica* V.1 1013$^a$16–17; *Aristoteles Latinus* XXV.2, p. 84.

59. *Summa theologiae* II, ii, q. 1, a. 5; *ed. cit.* vol. 2, p. 7b.

60. *Ad Hebreos* xi.1.

61. Godofridus de Fontibus, *Quodlibetum* VIII, q. 7; *ed. cit.* p. 73.

62. Durand de S. Pourçain, *In Petri Lombardi Sententias Theologicas Commentariorum libri IIII*, Prol. q. 1, § 36 (Venice: Typographia Guerraea 1571; reprt. Ridgewood, NJ: Gregg Press 1964), fol. 4$^{rb}$.

63. Landulfus Caracciolo, *Super I Sententiarum* Prol. q. 4, a. 1; Vienna, Nationalbibliothek 1496, fol. 7$^{vb}$.

64. Girardus de Bononia, *Summa theologiae* q. 1, a. 1, "Utrum theologia sit scientia"; Vatican, Borgh. lat. 27, fol. 1$^{ra}$-2$^{rb}$ at 2$^{rb}$.

65. I have been unable to identify this author, despite the relatively precise citation of the work. The majority of Antonius's citations come from Parisian theologians before 1330, but no record of a quodlibetal question by this author appears in Glorieux, *La littérature quodlibétique de 1260 à 1320* 2 vols. (Le Saulchoir: Kain, 1925; Paris: Vrin 1935). Likewise, there is no record of such a scholar in Glorieux's *Répertoire des Maîtres en théologie de Paris au XIII$^e$ siècle* 2 vols. (Paris:Vrin 1933–34) or *La faculté des arts et ses maîtres au XIII$^e$ siècle* (Paris: Vrin 1971). Casting the net somewhat wider, he cannot be found in the volumes of the *Chartularium Universitatis Parisiensis* before the time of Antonius, nor in the *Chartularium studii Bononiensis*, where Antonius might also have encountered him.

It is, of course, possible, that Antonius erred in identifying the author. MS R reads "May$^i$," which might be taken for "Mayronis"; this possibility proves equally fruitless, and were the reference to Franciscus de Mayronis, Antonius's frequent source, a search of Franciscus's quodlibetal questions fails to locate a suitable source for this reference. Short of an exhaustive search of university records, this individual at present must remain unidentified.

66. Antonius de Carlenis, *Questiones* . . . Prol. q. 2; edn. pp. 23, 26.

67. *Summa theologiae* I, q. 1, a. 2.

68. Girardus de Bononia, *Summa theologiae*, q. 2, a. 2, "Utrum theologia sit certissima scientiarum"; Vatican, Borgh. lat. 27, fol. 6$^{vb}$-7$^{vb}$ at 7$^{vb}$.

69. I *Ad Corinthios* xiii.10.

## \<QUESTIO 2:
## 'UTRUM THEOLOGIA SIT
## SUBALTERNATA VEL SUBALTERNANS'>*

Queritur[†] utrum theologia sit scientia[‡] subalternata vel subal-
5   ternans, et arguitur quod non, quia scientia Dei non potest esse
nisi una. Ergo non potest scientia divina subalternare[§] theologiam. // (V fol. 107va)

Secundo, scientia subalternata et subalternans non sunt primo
de eisdem veritatibus; sed eque potest esse de eisdem veritatibus,
10   de quibus est scientia beatorum.

Tertio, scientia subalternata non potest stare cum fide, sed ista
scientia, ut dicunt ponentes eandem conclusionem, stat cum fide.
Ergo non potest dici etiam[#] scientia subalternata.[**]

Quarto, habens scientiam subalternantem, potest habere sub-
15   alternatam. Sed viator non potest habere scientiam subalternantem cum subalternata. Ergo etc.[††]

Quinto, si hec scientia subalternaretur scientie beatorum, hoc
esset[‡‡] quia illa est clarior,[§§] et[##] ista obscurior,[***] et sic
secundum[†††] hoc, scientia[‡‡‡] notitie noctue subalternaretur sci-
20   entie aquile,[§§§] et // (O fol. 174rb) quelibet scientia minus clara
subalternaretur scientie clariori.

Sexto, scientia subalternata accipit notitiam eorum que[###]
probantur in scientia superiori, quod non est verum de scientia
theologie respectu scientie beatorum.[****]
25   Septimo, scientia subalternata semper habet subiectum com-

---

*M fol. 2va; O fol. 174ra; R fol. 2vb; V fol. 107rb.
†*Marg.:* "Vide in quarto notando in questione de subiecto." M.
‡scientia *om.* O, R.
§subalternare M, O, V ] subalternam R.
#etiam *om.* M, V.
**subalternata M, O, V ] subalterna R.
††Ergo etc. *om.* O, R.
‡‡esset M, R, V ] est O.
§§illa est clarior M, O ] clarior est illa R.
##illa est clarior et *om.* V.
***ista obscurior *transp.* R.
†††secundum *om.* M.
‡‡‡non esset *add. sed canc.* V.
§§§scientie aquile *transp.* R.
###que M, V ] qui O, R.
****scientie beatorum *transp.* R.

# <QUESTION 2:
## WHETHER THEOLOGY IS A
## SUBALTERNATE OR SUBALTERNATING SCIENCE>

The question is whether theology is a subalternate or subalternating science, and the argument is that it is not, since there can be only one science of God. Therefore divine science cannot subalternate [to itself] theology.

Second, a subalternate and a subalternating science are not concerned with the same truths primarily. But both can be concerned with the same truths, which pertain to the science of the blessed.

Third, a subalternate science cannot be compatible with faith. But this science is compatible with faith, as those holding this conclusion state. Therefore, it cannot be called a subalternate science.

Fourth, one who has the subalternating science can have the subalternate science. But the wayfarer cannot have the subalternating science along with the subalternate. Therefore.

Fifth, if this science were subalternated to the science of the blessed, it would be because the latter is clearer and the former more obscure. According to this, the knowledge of the owl would be subalternated to the knowledge of the eagle, and any science that is less clear would be subalternated to a science that is more clear.

Sixth, the subalternate science takes its knowledge from matters that are proved in the superior science. This is not true of the science of theology as related to the science of the blessed.

Seventh, the subalternate science always has a subject com-

positum ex duobus per accidens, sicut patet de perspectiva et
armonica.*

Octo, omnis scientia subalternata subalternatur duabus scientiis. Sed hec non† est huiusmodi. Ergo.‡ Probatur maior: cum su-
5   biectum componatur ex rebus diversarum rationum, habet
passiones // (M fol. 2vb) complexas,§ et per consequens principia
quibus ostenduntur passiones de subiectis, sicut patet de hoc //
(R fol. 3ra) principio, 'visio super assensam rectam est perfectissima.'# Una pars istius probatur in scientia naturali per illud
10  principium, 'agens** naturale magis approximatur,†† perfectius
agit.'

Nono, scientia subalternata respectu subalternantis‡‡ habet rationem causati, sed hoc non habet scientia viatorum respectu scientie beatorum, quia non dependet nisi ex subiecto vel§§ obiecto
15  vel lumine, et nullum illorum habet scientia beatorum. Ergo
etc.##

In contrarium, est illud quod ponit sanctus doctor hic.[1]

### <Quattuor notanda>

Primo, notandum quod Herveus[2] sequitur doctrinam sancti
20  doctoris, in questionibus ubi supra dicit quod duplex est subalternatio, // (V fol. 107vb) sicut*** dependentia videlicet ex parte
scibilis vel ex parte scientie, et dicit[3] quod 'theologia non subalternatur scientie beatorum subalternatione††† que est ex parte
scibilis, quia idem‡‡‡ sub eadem ratione formali est utrobique, ut
25  nunc suppono, nec est ibi dependentia ex parte cognoscentis
proveniens§§§ ex tali dispositione, que consequitur hoc individ-
uum vel illud, sed solum est // (O fol. 174va) ibi dependentia ex
parte cognoscentis, que provenit ex dispositione que per se et
naturaliter consequitur aliquem statum generalem hominum,'
30  quia manente tali modo naturali### cognoscendi quem nos habemus illo modo addito, oportet nos dependere in cognitione eorum que fides tradit, et**** de quibus theologia tractat. Et

---

*armonica M, R, V ] arithmetica O.
†non *om.* M, V.
‡Ergo *add.* M, V.
§passiones complexas *transp.* V.
#perfectissima *corr.* M.
**agens *om.* R.
††approximatur M, V ] approximatum O, R.
‡‡subalternantis M, O, V ] alternantis R.
§§ex *add.* R.
##Ergo etc. *om.* O.
***et *add.* O, R.
†††subalternatione M, O, V ] subalternationem R.
‡‡‡idem O, R ] vel M, V.
§§§proveniens M, V ] perveniens O, R.
###naturali M, O, R ] generali V.
****et *redupl.* R.

posed accidentally of two formalities, as is clear in the case of perspective and music.

Eighth, every subalternate science is subalternated to two sciences. But this science is not like this. Therefore. Proof of the major premise: [because] when a subject is composed of things having diverse formalities, it has properties that are complex and, consequently, corresponding principles by which the properties are demonstrated of the subjects, as is clear of this principle, "Vision is most perfect in a direct path." One part of this is proved in natural science through this principle, "The closer a natural agent, the more perfectly it acts."

Ninth, the relationship between the subalternate and subalternating science is causal, but this does not apply to the science of the wayfarer with respect to the science of the blessed, since [the former] depends only on the subject, or object, or the light [under which it is seen], and the science of the blessed has none of these. Therefore.

To the contrary, there is the position St. Thomas establishes here.

<Four Notations>

Note first that Hervé follows the teaching of St. Thomas in the question where he says above that subalternation is twofold, depending on whether it is on the part of the thing known or on the part of the science. And he says that "theology is not subalternated to the science of the blessed by a subalternation on the part of the thing known, since the same thing is known under the same formality in both, as I now assume, nor is there a dependence on the part of a knower that flows from such a disposition as is found in this or that individual; rather, the dependence is only on the part of a knower that comes about from a disposition that follows essentially and naturally on some general condition of men," since leaving intact our natural mode of knowing, there is an additional mode on which we must depend for knowledge of matters that pertain to faith and of which theology treats. And

dicit Herveus in fine[4] quod 'quando dico theologiam subalternari
scientie beatorum, non intelligo quod subalternetur ei sicut scientia subalternetur* scientie, sed subalternatur ei sicut habitus
creditivus habitui evidenti. Unde† unam scientiam subalternari
5    alteri ex parte scientis, non intellexi‡ hoc accipiendo proprie et
stricto modo scientiam,§ sed solum accipiendo largo modo scientiam, pro habitu cognitivo communiter dicto.'

Secundo, notandum quod Scotus in hoc prohemio[5] ponit opinionem sancti doctoris, et eam reprobat argumentis prepositis, et
10   Landulfus[6] similiter# argumentis posterius positis, qui sequitur
etiam** in hoc Petrum Aureoli[7] et Scotum, sed sicut dicit Egidius[8]
et bene hic in prohemio, in secundo argumento secunde questionis, nullus modus subalternationis qui conspicitur in humanis
competit isti subalternationi, quia // (R fol. 3rb) primus modus
15   subalternandi est quando aliqua scientia est de aliquo optimo,
alia vero non. Et isto modo non competit, quia tam scientia††
beatorum, quam theologia nostra principaliter est de Deo. Et sic‡‡
etiam alius modus subalternationis, qui est quia subalterna dicit
propter quid de eo de§§ quo subalternata dicit## quia, non com-
20   petit, ut ipse Egidius dicit, quia in Deo omnia sunt Deus, // (O fol.
174vb) et causa carent. Et sic etiam alius modus // (M fol. 3ra)
positus ab Aristotele hic non habet locum; tamen concedit Egidius quod largo modo, potest dici subalternata. Et habet similitudinem, cum unoquoque dictorum. Nam cum primo habet
25   similitudinem,*** quia scientia beatorum subalternat sibi theologiam, et licet utraque sit de optimo, illa tamen illud optimum
magis apprehendit, et etiam habet similitudinem cum alio modo,
quia principia non sunt clara in theologia; sunt tamen clara in scientia beatorum. Habet etiam††† similitudinem cum tertio modo
30   posito ab‡‡‡ Aristotele, quia sicut ea§§§ que tradit apparentia sive
navalis de cursu astrorum grosso modo, et

---

*subalternetur R ] subalternatur M, O, V2.
†unde O, R, V ] unam M.
‡intellexi M, R, V ] intelligi O.
§scientiam *om.* M, V.
#similiter *om.* O, R.
**sequitur etiam *transp.* M, V.
††quam *add. sed canc.* R.
‡‡sic M, O, V ] sicut R.
§§de eo de O, R, V ] de eodem M.
##subalternata dicit *transp.* M, V.
***cum unoquoque dictorum. Nam cum primo habet similitudinem, *om.* O.
†††Habet etiam *transp.* R.
‡‡‡ab *om.* M.
§§§ea M, O, R ] illa V.

Hervé says in the conclusion that "when I say that theology is subalternated to the science of the blessed, I do not mean that it is subalternated to it as one science to another, but as a habit of belief is subalternated to habit of evidence. Whence I do not mean that one science is subalternated to another on the part of the knower, taking science in a proper and strict sense, but rather taking it in a broad sense, for a cognitive habit understood in a general way."

Second, note that in this prologue Scotus presents the opinion of St. Thomas and argues against it with the foregoing arguments, and similarly Landulf [argues against it] with the subsequent arguments; in this, he follows Peter Aureoli and Scotus, but just as Giles says, indeed here in the prologue in the second argument of the second question, no type of subalternation that is found in human sciences is appropriate to this subalternation, since the first type of subalternation is when one science concerns an optimal subject, whereas the other does not. And this type is not appropriate, since the science of the blessed, just like our theology, is concerned principally with God. And so also the other type of subalternation—where the subalternating [science] gives the reasoned fact for what the subalternate gives only the fact—is not appropriate, as Giles himself says, since in God all things are God, and they all lack a cause. And so also the other type presented by Aristotle does not apply here; nevertheless, Giles concedes that in a broad sense it can be referred to as subalternate. It has a similarity with each of the aforesaid [types]. For it has a similarity with the first, in that the science of the blessed subalternates [our] theology to it, and although both are concerned with an optimal subject, nevertheless [the science of the blessed] apprehends that subject to a greater degree. It also has a similarity with the other type, since the principles are not clear in [our] theology whereas they are clear in the science of the blessed. There is even a similarity with a third type suggested by Aristotle, since just as the appearances which the navigator grasps about the paths of the stars in a more inexact way

astronomus* videt subtili modo, ita et theologus[†] intuetur grosso
modo, quod Deus et beati tuentur[‡] subtiliter. Et hanc opinionem[§]
Egidii approbat Giraldus[9] hic, qui etiam ponit[#] dictum quod dicit
esse sancti doctoris, quod sit scientia, et reprobat. Et in eodem
5    articulo ponit dictum Petri Aureoli,[10] qui dicit quod nullus potest
habere scientiam subalternatam sine subalternante, et[**] reprobat, quia aliquis potest acquirere[††] notitiam de principiis scientie
subalterne via sensus, memorie, et experientie. Et concludit subalternationem esse in proposito, ut ponit Egidius.
10    Tertio, notandum quod Durandus[11] sequitur viam Scoti,
ponendo argumentum quod scientia subalterna et subalternata
sunt compossibiles semper in eodem subiecto, et Herveus respondet huic fundamento Durandi, concedendo quod 'in[‡‡] subalternatione que est ex parte scientis, ubi est dependentia non
15    scientie a scientia, sed fidei a scientia, // (O fol. 175ra) quia quod
unus videt, alius tantum credit, est incompossibilitas, quia includit claritatem et obscuritatem respectu eiusdem, et ideo oportet[§§]
quod sint in diversis, vel[##] quod hoc non sit simul in eodem subiecto, sed in diversis temporibus.' Sed licet // (R fol. 3va) dicat
20    ita*** // (V fol. 108rb) Herveus in I *Sententiarum*,[12] tamen Jeromus
dicit in *Epistula ad Paulinum*,[13] "discamus in terris, quorum scientia perseverat in celis," et[†††] ideo possumus concedere theologiam quam hic habemus remanere in patria, licet fides evacuetur
I[‡‡‡] *Ad Corrinthios* c. 13,[§§§][14] ut nota sancti doctoris II[a] II[e], questio
25    1,[15] licet Durandus[16] et alii, ut Scotus[17] et Landulfus,[18] dicant
quod fides evacuatur[###] quantum ad enigma et obscuritatem,
quia ponunt de ratione fidei quod**** non sit in evidentia,
sicut[††††] in superiori articulo positum est.[‡‡‡‡][19]

---

*astronomus O, V ] astronimus M, R.
[†]theologus O ] theologicus *corr.* V ] theologicus M, R.
[‡]tuentur O, R ] intuentur M, V.
[§]proprie *add.* O.
[#]hic *add. sed canc.* M.
**et *om.* O.
[††]acquirere M, O, V ] habere R.
[‡‡]in *om.* O ] loquendo de R.
[§§]et ideo *add. sed canc.* O.
[##]hoc *add.* O.
***ita] sic *corr. in marg.* O; ita dicat M, V.
[†††]et *om.* M, V.
[‡‡‡]I M ] II O, R, V.
[§§§]13 *om.* O, R, V.
[###]evacuatur M, O ] evacuetur R, V.
****quod *om.* O, R.
[††††]sicut *corr. in textu* R.
[‡‡‡‡]est *om.* M.

the astronomer sees in a more precise way, so also the theologian contemplates in an inexact way what God and the blessed contemplate in a more precise way. And Girard agrees with Giles's opinion, and presents the position that he says is St. Thomas's, that it is science, and argues against it. And in the same article he presents the position of Peter Aureoli, who says that there can be no subalternate science without a subalternating science, and he argues against this, since someone could acquire knowledge of the principles of the subalternate science by way of the senses, memory, and experience. And he concludes that there is subalternation in the proposed case, as Giles suggests.

Third, note that Durand follows the path of Scotus, posing the argument that the subalternating and subalternate sciences are always simultaneously possible in the same subject, and Hervé responds to Durand's assumption by conceding that "in the subalternation that is on the part of the knower, where there is a dependence not of science upon science, but of faith on science (since that which the one sees the other only believes), it is not simultaneously possible, since it brings together clarity and obscurity with respect to the same thing, and it is necessary that they be in different things, or that this should not be in the same subject simultaneously, but at different times." But while Hervé says this in the first book of the *Sentences*, Jerome says in the *Letter to Romans* "Let us learn on earth about that of which knowledge endures in heaven." And therefore we can concede that the theology which we have here will remain in heaven, although faith will be set aside (according to I *Corinthians* chapter 13), as was noted by St. Thomas, II$^a$, II$^e$, question 1, although Durand and others, like Scotus and Landulf, say that faith will be set aside with respect to enigma and obscurity, since they posit that faith does not rest on evidence, as has been established in the article above.

Quarto, notandum quod concludendo, Herveus, in secunda
questione de questionibus factis ad petitionem magistri
Hemmerici,[20] dicit[21] quod 'videtur mihi quod quando Frater
Thomas scientiam subalternam vocat theologiam, quod ipsa non
5  habet rationem scientie proprie dicte, nisi inquantum continuatur
scientie // (M fol. 3rb) subalternanti, ideo idem Frater Thomas
quandoque dicit, quod nec fides nec theologia est de rebus scitis,
et quandoque vocat theologiam* subalternam, utrumque est
verum. Nam accipiendo scientiam proprie dictam, intelligitur
10 primum dictum; accipiendo autem scientiam subalternam largo
modo, pro omni habitu accipiente sua principia ab alio, dicit
theologiam esse scientiam† subalternam, et verum est.' Hec dicit
Herveus. Et idem in secunda questione‡ articulo primo circa
finem,[22] magis // (O fol. 175rb) clare hoc dicit, dicendo quod 'vi-
15 detur mihi, quando Frater Thomas in aliquo loco§ dicit theologiam esse scientiam subalternam, non dicit quantum ad hoc,
quod scientia subalterna inventa ab homine habet processum scientificum, sed quantum ad hoc, quod habet similitudinem cum
ea in hoc, quod sicut scientia subalterna# proprie dicta ab hom-
20 ine inventa habet** sua principia, saltem quantum ad propter
quid est†† ut credita sunt,‡‡ ita§§ etiam theologia habet sua principia credita. Sed non solum quantum ad propter quid est,## sed
etiam quantum ad quia est. Unde in I Sententiarum,[23] videtur
dicere quod theologia // (V fol. 108va) non est proprie, sed largo
25 modo dicitur scientia subalterna.' Hoc dicit*** Herveus.

---

*theologiam M, O, R ] scientiam V.
†scientiam om. R.
‡in secunda questione M, O, R ] in II^a, II^e, q. 5 V
§loco om. O.
#sicut scientia subalterna M, O, V ] scientia sub- sicut alternam R.
**habet om. M, V.
††est om. O, R.
‡‡sunt om. M, V.
§§ita M, R, V ] sic O.
##est om. O, R.
***Hoc dicit M, V ] Hec O, R.

Fourth, note that in concluding the second of the questions in response to Master Aimericus, Hervé says that "it seems to me that when Brother Thomas calls theology a subalternate science, it does not have the formality of science properly speaking, except insofar as it is in continuity with the subalternating science. Brother Thomas sometimes says the same, that neither faith nor theology concerns matters that are known, and sometimes he calls theology a subalternate science, both of which are true. For taking science in its proper sense, the first makes sense; but taking subalternate science in a broad sense, for every habit that takes its principles from another [discipline], it means that theology is a subalternate science, and this is true." That is what Hervé says. And in the second question, article one, near the end, he says the same thing more clearly, that "it seems to me, when Brother Thomas says elsewhere that theology is a subalternate science, he does not do so in the sense that a subalternate science created by man follows a scientific procedure, but in the sense that it is similar in that just as the subalternate science created by man properly speaking holds its principles (at least with regard to the reasoned fact) as believed, so also theology holds its own believed principles. But this is true not only with respect to the reasoned fact, but also with respect to the fact. Whence in the first book of the *Sentences*, he seems to say that theology is a subalternate science not properly, but broadly speaking." Hervé says this.

### *\<Alique conclusiones\>*

Prima conclusio: scientia theologie non* subalternatur methaphysice. Probatur, quia principia istius nullo modo declarentur[†]
in methaphysicalibus,[‡] cum sint[§] solum revelabilia solo[#] lumine
5  divino. Tamen negari non potest quin subiectum istius sit sub subiecto illius subiectatione predicationis, sicut magis explicabitur
in sequenti questione.[24] // (R fol. 3vb)

Secunda conclusio:** ista scientia non subalternat sibi alias hu-
manas. Probatur, quia alie non accipiunt principia ab ista,[††] licet
10  ancillentur isti secundum modum loquendi sancti doctoris[25] et
Egidii.[26] Et si arguitur contra, quia theologia est in[‡‡] cognitione
Dei, qui est omnium rerum causa, et sic videtur dicere propter
quid, respondet[§§] Egidius,[##27] quia hoc non sufficit, nisi illam
causam tradat secundum rationis modum, et non secundum rev-
15  elationis formam.

Tertia conclusio: tam ratione principiorum scientie theologice,
quam ratione clari luminis scientia beatorum subalternat sibi scientiam viatorum.*** Probatur ista conclusio ex dictis, quia clarum
est quod subiectum est idem, sed quantum ad evidentiam
20  dictorum,[†††] ista scientia subalternatur[‡‡‡] illi.

Quarta conclusio. Patet[§§§] ex supradictis[28] quod moderate est
intelligenda.

---

*non *om.* R.
[†]declarentur O, R ] declarantur M, V.
[‡]methaphysicalibus M, O, V ] methaphysice R.
[§]sint M, V ] sicut O, R.
[#]solo M, O, R ] suo V.
**M *marg.*: "Nam bene conceditur quod theologia secundum modum suum quia pro-
cedit ex principiis nobilissimis ut puta articulis fidei quantum ad veritates divinas etc.,
quod sit dignissima et nobilissima scientiarum, sed non quod sit vera scientia ut dicunt
multi doctores."
[††]ista M, V ] illa quia R ] illa O.
[‡‡]in M, V ] de R ] *om.* O.
[§§]respondet M, O, V ] et R.
[##]M *Marg.*: "Nota."
***viatorum M, R, V ] viatoris O.
[†††]dictorum M, R, V ] viatoris O.
[‡‡‡]subalternatur O, R ] subalternetur M, V.
[§§§]quia *add.* M.

*<Some Conclusions>*

The first conclusion: the science of theology is not subalternated to metaphysics. Proof: its principles are in no way formulated in metaphysical terms, since they are revealable only by divine illumination. Nevertheless, it cannot be denied that the subject of the latter [science] falls under the subject of the former, by a subalternation of predication, as will be explained in greater detail in the following question.

The second conclusion: this science does not subalternate other human [sciences] to itself. Proof: the others do not take their principles from it, though they are its handmaidens, following St. Thomas's and Giles's terminology. And if one should argue against this, that theology consists in the cognition of God, who is the cause of all things, and so [theology] would seem to provide the reasoned fact, Giles responds that this does not suffice [as an explanation] unless it treats that cause according to the mode of reason, and not according to the form of revelation.

The third conclusion: the science of the blessed subalternates the science of the wayfarer to itself both by reason of the principles of theological science and by reason of the clear light [of revelation]. Proof of this conclusion: from what has been said previously, since it is clear that the subject is the same, but with regard to the evidence of the things said, the latter science is subalternated to the former.

The fourth conclusion: it is clear from the things said previously that this must be understood properly.

## \<*Ad argumenta principalia*\>

Ad primum argumentum,[*] quod ymmo est alia scientia nostra // (O fol. 175va) a scientia Dei et beatorum, licet fiat argumentum in contrarium, quia idem est subiectum, scilicet Deus, et sub ea-
5  dem ratione formali, deitatis. Sed hoc[†] solvetur infra.[29]

Ad secundum, quod non sunt omnino primo de eisdem[‡] veritatibus scientia beatorum et scientia viatorum.[§] // (M fol. 3va)

Ad tertium, quod scientia de eodem et fides non possunt esse secundum doctrinam sancti doctoris, licet Durandus,[#][30]
10  Scotus,[31] et Landulfus[32] et sequentes eos dicant quod fides et scientia possunt esse de eodem, sed scientia theologie, eo[**] modo quo habet rationem scientie, non habet rationem fidei, quia fides infunditur, ista acquiritur; et si iste habitus dicatur creditivus, aliquo modo potest etiam dici scientificus, ut declaratum est
15  supra.[††][33]

Ad quartum,[‡‡] ut[§§] patet ex secundo notando,[34] ista scientia non proprie est[##] subalternata, et sic[***] potest responderi secundum Herveum,[35] quia 'in subalternatione que est ex parte scientis, ubi est dependentia, non scientie a scientia, sed fidei a
20  scientia, est incompossibilitas.' Sed potest responderi adhuc alio modo, quod ut dictum est in quarto notando,[36] beati retinent habitum theologicum.

Ad quintum, quod facit Scotus[37] et Landulfus,[38] dicendum est quod cognitio clara subalternat sibi minus claram quando minus
25  clara dependet ab ipsa, sicut supradictum[†††] est,[39] non sic est de visione aquile et noctue. Et sic[‡‡‡] nullo modo est ad propositum.

---

[*]argumentum *om.* O, R.
[†]hoc M, O, V ] hic R.
[‡]eisdem M, O, V ] eiusdem R.
[§]viatorum M, R, V ] viatoris O.
[#]et *add.* M.
[**]eo O, R ] eodem M, V.
[††]supra M, R, V ] prius O.
[‡‡]quod *add.* M, V.
[§§]ut. *om.* R.
[##]proprie est *transp.* M, V.
[***]sic M, O, V] si R.
[†††]supradictum R ] predictum O ] superius dictum M, V.
[‡‡‡]Et sic O, R ] Unde M, V.

*<Responses to the Principal Arguments>*

To the first argument: our science is indeed different from the science of God and the blessed, though an argument may be made to the contrary, that it has the same subject, God, and the same formal consideration, that of deity. But this will be resolved below.

To the second, the science of the blessed and that of the wayfarer are not about the same truths altogether primarily.

To the third, according to the teaching of St. Thomas, science and faith cannot be of the same object, although Durand, Scotus, and Landulf and those following them say that faith and science can be of the same object, for the way in which the science of theology has the formality of science it does not have the formality of faith, for faith is infused whereas science is acquired. And if this habit should be said to be believed [*creditivus*], in a certain way it can also be said to be scientific [*scientificus*], as has been declared above.

To the fourth: as is clear from the second notation, this science is not subalternated properly speaking, and so one can respond in the way that Hervé does, that "in the subalternation that is on the part of the knower, where there is a dependence not of science on science, but of faith on science, this is not possible." But one could still respond in another way that, as has been said in the fourth notation, the blessed retain the habit of theology.

To the fifth, the argument that Scotus and Landulf make: it must be said that the knowledge that is more clear subalternates the less clear to itself when the less clear depends upon it, as has been said above; this does not happen with the vision of the eagle and owl. And so this is in no way pertinent to the issue.

Ad sextum, quod hec scientia accipit notitiam eorum que clara sunt in scientia superiori beatorum, qui sciunt causas et propter quid.

5 Ad septimum, quod est Landulfi,[40] potest simpliciter negari quod semper scientia subalternata* // (R fol. 4ra) habet subiectum compositum ex duobus[†] per accidens, quia secundum plures, scientia physicorum[‡] subalternatur methaphysice, et tamen non subalternatur alteri. Secundo,[§] potest responderi quod illud exemplum[#] quod dat est** de scientiis partim naturalibus et par-
10 tim mathematicis,[††] in tantum quod in II *Physicorum*[41] sanctus doctor sentit quod sint magis naturales, et Egidius[42] tenet quod sint magis mathematice,[‡‡] et potes addere intentionem Hervei in I *Sententiarum*,[43] quod hec scientia subalternatur scientie beatorum, non ex parte scibilis, sed ex parte scientis.

15 Ad octavum, quia est idem cum tertio, patet eadem responsio, quia nunc[§§] subiectum est compositum ex duobus quando subalternatur duabus scientiis; et probatio quam facit[##] de illo principio, 'visio super assensam rectam est perfectissima,' etc.,[***] hoc declarat.

20 Ad nonum, quod hec scientia habet[†††] rationem causati et dependentis[‡‡‡] respectu luminis in subiecto viatoris exeuntis,[§§§] et etiam potest dici respectu aliorum, ut dicit Egidius, et[###] dictum est in secundo notando.[44]

---

*semper scientia subalternata M, O, R ] scientia subalternata semper V.
†duobus M, O, V ] rationibus R.
‡physicorum M, O, V ] philosophicarum R.
§M *marg.*: "Sed posset etiam responderi, quia dato quod subiectum alicuius scientie sit compositum ex duobus, sed quia ex illis aggregatis resultatur quidam conceptus simplex, non oportet ipsam scientiam duabus subalternari, et hoc etiam docet iste doctor infra, questionibus libri *Posteriorum*, quare etc."
#illud exemplum *transp.* O, R.
**est *om.* R.
††mathematicis M] methaphysicis O, R, V.
‡‡mathematice M, O ] methaphysice R, V.
§§nunc M, O, V ] tunc R.
##facit O, R ] fecit M, V.
***etc. *om.* O, R.
†††habet *om.* M, V.
‡‡‡dependentis M, O, V ] dependenti R.
§§§exeuntis O, R ] existentis M, V.
###ut *add.* R.

To the sixth, this science receives knowledge of matters that are clear in the superior science of the blessed, who know the causes and the reasons why.

To the seventh, Landulf's argument: one can simply deny that the subalternate science always has a subject composed accidentally of two formalities, since according to many, the science of physics is subalternated to metaphysics, and nevertheless it is not subalternated under another formality. Second, one can reply that the example he gives concerns sciences that are partly physical and partly mathematical, in the sense that St. Thomas thinks in the second book of the *Physics* that they are more physical and Giles holds that they are more mathematical. You could add [to this] the position of Hervé in the first book of the *Sentences*, that this science is subalternated to the science of the blessed, not on the part of the thing known, but on the part of the knower.

To the eighth: since it is the same as the third, the same response will suffice, since now the subject is composed of two formalities, being here subalternated to two sciences, and the proof which he gives of this principle, "Vision in a straight line is most perfect," declares this.

To the ninth: this science has the formality of being caused and being dependent with respect to the light going forth into the subject of the wayfarer, and the same could also be said with respect to others, as Giles says and as has been expressed in the second notation.

## Notes

1. *Summa theologiae* I, q. 1, a. 2; *ed. cit.* vol. 1, p. 3b.

2. Hervé Natalis, *Defensa doctrinae sancti Thomae* prima pars, II, 2., a. 10; ed. Engelbert Krebs, *Theologie und Wissenschaft nach der Lehre der Hochscholastik* (Münster i. W.: Aschendorffsche Verlagsbuchhandlung 1912), 49*, n. 1.

3. Hervé, *Defensa doctrinae* prima pars, II, 2., a. 10; ed. Krebs, 50*.

4. Hervé, *Defensa doctrinae* prima pars, II, 2., a. 10; ed. Krebs, 50*.

5. Scotus, *Reportata Parisiensia* Prologue, q. 2; *Opera Omnia* vol. 22 (Paris: Vivès 1894), 34–37.

6. Landulfus Caracciolo, *Super I Sententiarum* Prologue, q. 2, a. 3; Vienna, Nationalbibliothek 1496, fol. 5$^{ra}$.

7. Petrus Aureoli, *Scriptum super primum Sententiarum* Prooem., sect. 1 B, art. 1a, §24–45; ed. Eligius Buytaert, volume 1 (St. Bonaventure, NY: Franciscan Institute 1952–1956), 139–145.

8. Egidius Romanus, *Primus Sententiarum* 2 princ., q. 1, a. 2; (Venice: Octavianus Scotus 1521), fol. 4$^{va}$K-L.

9. Girardus de Bononia, *Summa theologiae* q. 2, a. 3, "Utrum theologia sit subalterna"; Vatican, Borgh. lat. 27, fol. 7$^{va}$-10$^{vb}$ at 7$^{va-b}$.

10. Petrus Aureoli, *Scriptum* I, Prooem, sect. 1 B, art. 1a, § 36; ed. Buytaert, volume 1, p. 142.

11. Durand de S. Pourçain, *In Petri Lombardi Sententias Theologicas Commentariorum libri IIII* Prologue, q. 7, 4; (Venice: Typographia Guerraea 1571; reprt. Ridgewood, NJ: Gregg Press 1964), fol. 12$^{ra}$.

12. Hervé, *In IV libros Sententiarum* I, q. 6; (Paris: Dyonisius Moreau 1647), 15a-16a. Cf. *Defensa doctrinae* prima pars, II, 2., a. 10; ed. Krebs, 49*.

13. Jerome, *Epistola LIII. ad Paulinum* 10; ed. Isidorus Hilberg. CSEL 54 (Vienna-Leipzig: F. Tempsky, G. Freytag 1910), 464.

14. I *Ad Corinthios* xiii.12.

15. *Summa theologiae* II, ii, q. 1, a. 5; *ed. cit.* vol. 2, p. 7b.

16. Durand de S. Pourçain, *In Petri Lombardi Sententias* III, d. 31, q. 3, 11; *ed. cit.*, fol. 258$^{rb}$.

17. Scotus, *Reportata Parisiensia* III, d. 31, a. unic.; *Opera Omnia* (Paris: Vivès 1894), 500b–501a.

18. Landulfus Caracciolo, *Super I Sententiarum* Prologue, q. 4, a. 1; Vienna, Nationalbibliothek 1496 fol. 7$^{vb}$.

19. *Supra*, q. 1; edn. pp. 7–9.

20. Aimericus de Placentia, magister generalis Ordinis Praedicatorum (1304–1311). See Thomas Kaeppeli, *Scriptores ordinis Praedicatorum Medii Aevi* 1 (Rome: S. Sabina 1970), 19–21.

21. Hervé Natalis, *Defensa doctrinae* prima pars, I. a. 3; ed. Krebs, pp. 10*–11*. Cf. the complete text in Vatican, lat. 817, fol. 3$^{vb}$.

22. Hervé Natalis, *Defensa doctrinae* prima pars, II, 1., a. 7; ed. Krebs, 37*.

23. Aquinas, *Scriptum super libros Sententiarum* Prologue, q. 1, a. 3; ed. P. Mandonnet vol. 1 (Paris: P. Lethielleux 1910), 13. But see also Krebs's remarks, *Theologie und Wissenschaft*, 37*, n. 1.

24. *Infra*, q. 4, prima conclusio; Rome, B. Casanatense 1025, fol. 10$^{vb}$.

25. *Summa theologiae* I, q. 1, a. 5 sed contra; *ed. cit.* vol. 1, p. 5a.

26. Egidius Romanus, *Primus Sententiarum* 2 princ., q. 1, a. 3; *ed. cit.* fol. 4$^{vb}$P.

27. Ibid.

28. *Supra*, esp. 24.

29. *Infra*, q. 4: "Utrum Deus sit subiectum in theologia"; Rome, B. Casanatense, 1025, fol. 8$^{ra}$-12$^{vb}$.

30. Durand de S. Pourçain, *In Petri Lombardi Sententias* Prologue, q. 1, § 36; *ed. cit.* fol. 4$^{rb}$.

31. Scotus, *Reportata Parisiensia* III, d. 24, q. unica; *Opera Omnia* vol. 23 (Paris: Vivès 1894), 453–454.

32. Landulfus Caracciolo, *Super I Sententiarum* Prologue, q. 4, a. 1; Vienna, Nationalbibliothek 1496, fol. 7$^{va}$.

33. *Supra*, 24.

34. *Supra*, 22–23.

35. Hervé, *In IV libros Sententiarum* I, q. 6; *ed. cit.*, p. 16a.

36. *Forsitan* tertio notando; *supra*, 23.

37. Scotus, *Ordinatio* Prologue, pars 4, q. 1–2, 216; *Opera Omnia* vol. I (Vatican: Polyglottis 1950), 147–148.

38. Landulfus, *Super I Sententiarum* Prologue, q. 4, a. 1; Vienna, NB 1496, fol. 7$^{va}$.

39. *Supra*, 22–23.

40. Landulfus, *Super I Sententiarum* Prologue, q. 2, a. 4; Vienna, NB 1496, fol. 5$^{rb}$.

41. Aquinas, *In VIII libros physicorum Aristotelis Expositio* II, lect. 3, § 164; ed. P. M. Maggilo (Turin-Rome: Marietti 1954), 84.

42. Cf. Egidius Romanus, *In libros de physico auditu Aristotelis Commentaria* II, lect. 3 (Venice: Octavianus Scotus 1502), fol. 29$^{ra}$, 31$^{va}$.

43. Hervé, *In IV libros Sententiarum* I, q. 6; *ed. cit.*, pp. 15b–16a.

44. Egidius Romanus, *Primus Sententiarum* 2 princ., q. 1, a. 2; *ed. cit.* fol. 4$^{va}$K-L. *Supra*, 22–23.

# ANTONIUS DE CARLENIS DE NEAPOLI
## *QUESTIONES IN LIBROS*
## *I-II ANALYTICORUM POSTERIORUM*
## *ARISTOTELIS*
## L. I, QQ. 17, 22

# Conspectus Siglorum

C = Chicago, Newberry Library, Case MS 97,5 (a. 1468)

· · · · ·

| | |
|---|---|
| *add.* | addidit |
| *canc.* | cancellavit |
| *conj.* | conjectura |
| *corr.* | correxit |
| *del.* | delevit |
| *expand.* | expandit |
| *ins.* | inseruit |
| *lin.* | linea(m) |
| *marg.* | margine |
| *dext.* | dextro |
| *sinist.* | sinistro |
| *obscur.* | obscuratus |
| *om.* | omisit |
| *ras.* | rasura |
| *suppl.* | supplevit |
| *transp.* | transposuit |
| < > | omissiones supplevimus |
| / . . . / | littera illegibilis |

## \<L. I, Q. 17:
## 'UTRUM DEMONSTRATIO POSSIT ESSE
## EX EXTRANEIS'>*

Decimo septimo, queritur utrum demonstratio possit esse ex
5   extraneis, et arguitur quod sic, primo per textum huius, quia non
sit mentio nisi de subiecto et passione et dignitate, quod debent
esse principia propria. Ergo non intelligitur de premissis propriis
que in qualibet sciri non dicuntur dignitates.

Secundo, scientie possunt descendere de genere in genus.
10   Ergo non oportet quod quelibet scientia procedat ex propriis et
non extraneis. Probatur consequentia, quia si licet fateri descen-
sum, sequitur quod ista scientia, in qua descenditur de genere in
genus, procedat ex extraneis. Antecedens autem probatur, quia
methaphysica descendit in alias scientias et probat principia
15   earum, et dyalectica docet artem silogizandi et habet principia
aliarum scientiarum; similiter astrologus considerat de eclipsi et
cum eclipsis sit, quid naturale considerare de ipsa partium ad
naturalem. Ergo in astrologiam descendit scientia naturalis, cum
astrologia utatur hiis que ad naturalem pertinent.

20   Tertio, arguitur per textum,[1] quia si† proceditur ex extraneis,
tunc subiectum, passio, et medium non erunt eiusdem generis.
Sed hoc est falsum, quia ista non sunt eiusdem // (fol. 18va) generis, cum medium sit quoddam complexum, et subiectum erit
substantia, et passio accidens.

25   Quarto, saltem in scientia methaphysica illud esset falsum,
quia nec subiectum nec passio sunt unius generis, cum sint
transcendentia.

Quinto, idem est procedere non ex propriis et procedere ex extraneis; sed methaphysica procedit non ex propriis, cum semper
30   procedat ex communibus.

---

*C fol. 18rb.
†non *add. sed canc.* C.

## \<BOOK I, QUESTION 17: "WHETHER DEMONSTRATION CAN BE MADE FROM EXTRANEOUS PREMISES"\>

Seventeenth, the question is whether demonstration can be made from extraneous premises, and the argument is that it can, first, from Aristotle's text, since it mentions that the principles must be proper only with respect to the subject, the property, and the axiom. Therefore, there is no question here of proper premises, which are not called axioms in any particular science.

Second, sciences can descend from one genus into another. Therefore, it ought not be the case that any science should proceed from principles that are proper and not extraneous. Proof of the consequence: if a descent is permitted, it follows that any science that descends from one genus to another proceeds from extraneous principles. Proof of the antecedent: metaphysics descends into other sciences and proves their principles, and dialectic teaches the art of syllogizing and contains the principles of other sciences. Similarly, the astronomer considers an eclipse and its accompaniments; the natural philosopher considers the same according to the parts pertaining to natural philosophy. Therefore, the science of nature descends into astronomy, since astronomy uses matters that pertain to natural philosophy.

Third, an argument from the text: if it should proceed from extraneous principles, then the subject, the property and the middle term will not be of the same genus. But this is false, since these are not of the same genus, since the middle term is a complex entity, and the subject is a substance, and the property is an accident.

Fourth: this would be false, at least in the science of metaphysics, since neither the subject nor the property is of a single genus, since they are transcendentals.

Fifth: to proceed from principles that are not proper is the same as to proceed from extraneous principles; but metaphysics proceeds from principles that are not proper, since it always proceeds from common principles.

Et confirmatur*: sicut se habet res ad esse, ita ad cognosci. Sed ita est quod virtus solis, cum sit communis, sufficit ad generationem eorum que fiunt ex putrefactione. Ergo et principia communia que non erunt propria sufficiunt ad demonstrationem,
5    aliter poneri principia communia videtur superfluum.

Sexto, per textum[2] huius videtur quod arismetica descendit in geometriam econtra, quia dicendo quod duo numeri cubi faciunt unum cubum, cubus pertinet ad geometriam, et numerus ad arismeticam, et tamen sunt scientie distincte. Ergo et principia
10   erunt extranea in illis scientiis.

Septimo, geometria descendit in cirurgiam, saltem quoad illam partem qua procedit quod vulnera circularia tardius sanantur vulneribus angularibus.[3]

Octavo, quia philosophia descendit in medicinam, cum subiec-
15   tum philosophie sit corpus mobile vel ens mobile, et subiectum medicine est corpus sanabile, quod est inferius ad illud, et tamen medicina est non pars philosophie naturalis ut de se patet.

In contrarium,[†] est Philosophus qui hic probat,[4] quod non est ex alio genere in aliud genus demonstrare, et ex hoc interponit
20   quomodo ex corruptibilibus non est demonstratio, et tandem concludit quod demonstratio non debet esse ex communibus, sed ex propriis ibi *Quoniam autem manifestum,*[5] et ex hoc declaratur que dicuntur esse principia communia ibi *Difficile autem,*[6] ibi *Dico autem principia in unoquoque genere.*[7]

25                   *<Alique notanda>*

Primo, notandum quod sanctus doctor[8] hic super textum, quod "non est descensus neque transitus de genere in genus," et si est unum genus distinctarum scientiarum, est "sic, id est, quodammodo. Aliter, enim, est impossibile quod demonstretur aliqua
30   conclusio ex principiis, nisi sit idem genus, vel simpliciter, vel secundum quid." Est simpliciter idem genus "quando ex parte subiecti non sumitur aliqua differentia determinans, que sit extranea a natura illius generis subiecti; sicut si quis per principia verificata

---

*Marg.:* "Confirmatio."
†*Marg.:* "In oppositum."

Confirmation: a thing is related to being as it is to knowing. Thus it is that the power of the sun, although common, suffices to generate things that come to be from putrefaction. Therefore principles that are common and not proper suffice for demonstration, for otherwise positing common principles would seem to be superfluous.

Sixth: from this text it seems that arithmetic descends into geometry and conversely, since in saying that two cube numbers make one cube, 'cube' pertains to geometry, and 'number' to arithmetic, and nevertheless these are distinct sciences. Therefore the principles will also be distinct in those sciences.

Seventh: geometry descends into surgery, at least with regard to the part that shows that circular wounds heal more slowly than those with angles [i.e. those that are jagged].

Eighth: philosophy descends into medicine, since the subject of philosophy is the changeable body or changeable being, and the subject of medicine is the curable body, which is contained under the former, and nevertheless medicine is not a part of natural philosophy, as is clear in itself.

To the contrary is [the text of] the Philosopher, who proves here that demonstration is not from one genus to another, and from this he interposes how there is no demonstration from corruptible things, and finally he concludes that demonstration must not be from common principles but from proper principles, at the passage "However, since it is evident. . . ." From this he explains which principles are said to be common at "It [is] difficult, however, . . ." [and] at "I call, however, principles in any genus. . . ."

### *\<Four Notations\>*

First, note what St. Thomas [says] here at this text, "There is no descent or movement from one genus to another," and if there were one genus of distinct sciences, it would be "so, that is, in a certain way. For otherwise, it is impossible that a conclusion be demonstrated from principles unless the genus is the same, either absolutely or in a qualified way." The genus is the same absolutely "when, with respect to the subject, no determining differentia is assumed that is extraneous to the nature of that subject genus; for example, if one attempted to demonstrate a

de triangulo procedat ad demonstrandum aliquid circa ysochelem, vel aliquam aliam speciem trianguli. Sed secundum quid est idem genus quando assumitur circa subiectum aliqua differentia extranea a natura illius generis subiecti, sicut visuale
5    extraneum est a genere linee, et sonorum extraneum est a genere numeri. Numerus ergo simpliciter, qui est genus subiectum arismetice, et numerus sonorum, qui est genus subiectum musice, non sunt unum genus simpliciter; similiter nec linea simpliciter, quam considerat geometra, et linea visualis, quam considerat per-
10   spectivus. Unde patet quod quando ea, que sunt vere linee simpliciter, applicantur ad lineam visualem, est quodammodo descensus in aliud genus, non autem quando ea que sunt trianguli applicantur ad ysochelem," qui est species trianguli. Et hic ad hoc inducit textus,[9] quod nulla scientia procedit ex principiis
15   alterius generis. Et hoc probat // (fol. 18vb) maxime in scientiis demonstrantis disparitis. Dat exemplum textus,[10] quod geometra non demonstrat quod duo cubi sunt unus cubus, de quo hic vide per sanctum doctorem[11] et alios expositores. Et ex hoc inducit Aristoteles[12] quod demonstratio non est ex corruptibilibus
20   propter quam causam; hic tractat propter textum quomodo diffinitio potest esse principium et conclusio, et potest etiam esse demonstratio solum positione differens, de quo etiam habes in II *Physicorum* in fine,[13] et infra in II huius,[14] et in VII *Methaphysice* in fine,[15] et hic etiam tractat sanctus doctor.[16] Aliqua propositio est
25   necessaria in ordine ad causam, de quibus omnibus supra diximus,[17] et sic vide ordinem textus, quia ibi *Quoniam autem manifestum est*,[18] declarat quod scientie non procedunt ex communibus principiis, licet communia principia cadant in scientiis secundum quandam proportionem sive analogiam,* ubi repre-
30   henditur Brisso,[19] quia voluit demonstrare quadraturam circuli per quedam communia, et sic concludit Aristoteles[20] declarande que sunt principia et communia quod discernere est valde difficile, ut supra discernimus.[21]

---

*analogiam ] anologiam C.

conclusion concerning an isosceles or other kind of triangle through a principle that is true of a triangle in general. The genus is the same in a qualified way when a differentia affecting the subject is taken that is extraneous to the nature of that subject genus, as 'visual' is extraneous to the genus of line and 'of sounds' to the genus of number. Therefore, number, which is the subject genus of arithmetic, and the number of sound, which is the subject genus of music, are not one genus absolutely; nor, similarly, are line absolutely, which the geometer considers, and visual line, which the perspectivist considers. Whence, it is clear that when conclusions that are true of a line absolutely are applied to a visual line, there is a kind of descent into another genus; this is not the case when conclusions that pertain to triangles are applied to isosceles," which is a species of triangle. And here the text introduces the point that no science proceeds from principles of another genus. He proves this especially for sciences that demonstrate different conclusions. The text gives the example that the geometer does not demonstrate that two cubes make one cube, concerning which see St. Thomas and other expositors. And from this Aristotle argues that demonstration is not based on corruptible things for the same reason. From the text here, he deals with the way a definition can be a principle and a conclusion, and can even be a demonstration, differing only in position. You will find a discussion of this at the end of *Physics* II, and below in the second book of this work, and at the end of *Metaphysics* VII, and in that place St. Thomas also treats it. A proposition is necessary in [its] relation to a cause, about which we spoke above; see the outline of the text, since at that place, "Since however it is evident . . . ," he declares that sciences do not proceed from common principles, although common principles fall within sciences according to a certain proportion or analogy. Here he reproves Brisso, who wished to demonstrate the quadrature of the circle through common principles. So Aristotle concludes by declaring that it is rather difficult to discern what principles are [proper] and what are common, as we have brought out above.

Secundo, notandum quod ex supra dictis, patet intentio textus, quia non licet descendere de genere in genus ex eo quia subiectum, passio et dignitas debent esse unius generis, idest unius proportionis, sicut etiam exponit hic sanctus doctor,[22] et alii expositores post eum. Et licet Aristoteles hic non faciat mentionem de premissis, tamen sub nomine dignitatis comprehenduntur, ut dicit Egidius,[23] et ratio* quia in extremis clauduntur media, et quod facit mentionem de extremis dat intelligere medium.

Sed hec responsio† non placet magistri Pauli‡,[24] quia ratione reductionis etiam non debuisset facere mentionem de dignitatibus, quia subiectum et passio resolvuntur in transcendens, et similiter arguit contra secundam, sed iste impugnationes sunt verbales; ideo pretermitto.

Tertio, notandum quod eadem ratione, scientie non debent esse ex communibus, sed ex propriis; unde non debent esse principia alia, et hic notat magister Paulus[25] quomodo aliquod dicitur extraneum dupliciter: primo, quia sibi non§ competit, sicut duos cubos faceri unum cubum est extraneum triangulo; secundo, quia competit alteri, et sic genus est extraneum speciei, et hoc modo principia communia sunt extranea cuilibet speciei specialissime, et licet genus ponatur in diffinitione speciei, et universaliter predicetur de ipsa, tamen non vocatur esse principium non extraneum, et sic aliquid potest esse commune intraneum et extraneum. Et ex hiis dictis, secuntur duo. Primum est quod demonstratio est ex principiis propriis; ergo non ex communibus tantum, et si ex propriis, non extraneis. Et etiam sequitur quod licet dicitur subalternatio scientiarum et subalternans descendat in subalternatam, tamen numquam scientia dicitur procedere ex principiis extraneis vel communibus precisis de qua subalternatione; licet hic tangatur aliquid, infra[26] faciemus questionem specialem.

---

*Marg.: "Quare Aristoteles non facit mentionem de dignitatibus."
†Marg.: "Responsio Egidii."
‡Marg.: "Improbatio magistri Pauli."
§non add. supra lin. MS.

Second, note from what has been said above that the intention of the text is clear, since descent from one genus into another is not permitted because the subject, property, and axiom must be of the same genus, that is, proportionally the same, as St. Thomas explains here, as do other expositors who follow him. And although Aristotle does not mention premises here, nevertheless they are included under the term 'axiom,' as Giles says. The reason is that the middle term is enclosed within the extreme terms, and the fact that he mentions the extremes means that he also refers to the middle term.

But this response does not please master Paul, since one must not refer to axioms in a reductive proof, since the subject and property are resolved by transcending [them], and he argues against the second [response] in a similar way. But these criticisms are merely verbal, and so I pass over them.

Third, note that by the same argument sciences should not proceed from common but from proper principles, and there must be no others. Master Paul observes at this point that something may be said to be extraneous in two ways. In the first way, because it is not suitable to its own [genus], just as two cubes making one cube is extraneous to triangle. In the second way, because it is suitable to something else, and so genus is extraneous to species, and in this way common principles are extraneous to any particular species. And although the genus is placed in the definition of the species and is predicated universally of it, nevertheless it is not said to be a non-extraneous principle. And so something can be intrinsically common and yet extrinsic. And from these conclusions, two others follow. The first is that demonstration is made from proper principles, and therefore not from those that are merely common; and if only from propers, not from those that are extraneous. It also follows that although one speaks of the subalternation of the sciences, and of the subalternating science descending into the subalternate science, a science is never said to proceed from extraneous principles or from principles that are common precisely in the way in which there is subalternation, although this touches on a matter we will consider in a special question below.

Quarto, notandum quod Aristoteles hic notat[27] que dicuntur hec esse principia communia, et que propria et distincta, quod propria in unoquoque genere quecumque sunt non contingit demonstrare, et dat exemplum de principiis communibus, ut si 5 ab equalibus equalia auferas, que remanent sunt equalia. Et de principiis propriis, dat exemplum; // (fol. 19ra) ibi patet etiam accipit principia propria pro subiectis propriis. Unde dicit de eis quod recipiuntur esse et hoc esse, id est, de eis presupponitur quid est et quia est; principia autem communia accipit quelibet 10 scientia particularis certa et limitata ad proprium genus, et hoc est dicere secundum analogiam, ut in textu dicitur.

<Quattuor conclusiones>

Prima conclusio: scientia debet esse ex propriis. Probatur ex textu,[28] et sic vides quia Aristoteles supra dixit[29] in declaratione 15 diffinitionis datis de demonstratione, ubi dixit quod debet esse ex primis, tandem concludit quod, "ex primis <autem est quod> ex principiis <propriis est>; idem enim dico primum et principium."[30] Et hoc declaratur, quia effectus requirit proprias causas, et determinati effectus determinatas causas, ut II 20 Physicorum,[31] et sic determinata demonstratio requirit determinatum principium.

Secunda conclusio: in demonstrando, non licet descendere de genere in genus. Probatur ex textu,[32] quia si sic, demonstratio non esset ex propriis, sed ibi descensus bene potest fieri de ge-25 nere predicato respectu alterius generis sibi subalternati, sicut respectu specierum. Nam a* mathematica, que† est ex duabus speciebus, potest abstrahi unus conceptus genericus. Et sic geometria, que est una scientia, procedit de genere in genus, cum descendit ad genera propinqua multarum suarum spe-30 cierum, ut est ysocheles, scalenem, etc., et sic etiam non capitur

---

*a add. supra lin. C.
†que add. supra lin. C.

Fourth, note that Aristotle discusses here what principles are said to be common and what are proper and distinct, and that one cannot use proper principles to demonstrate in any genus whatsoever. And he gives the example of common principles, that if you take equals from equals, the results are equal. And of proper principles he gives an example, from which it is clear that he there takes proper principles to mean proper subjects. Whence he says of things that receive 'being' [*esse*] and 'this being' [*hoc esse*], both what they are and that they are is presupposed. But any particular science takes common principles as certain and limited to a proper genus, and this is to speak analogically, as is said in the text.

<Four Conclusions>

The first conclusion: Science must proceed from proper principles. Proof, from the text; and so you see what Aristotle meant in the text cited above, in explaining the definition given for demonstration, where he said that it must proceed from premises that are primary, and at length he concluded that "[depending] on premises that are primary is the same as [depending] on proper principles; for I say that to be primary and to be a principle are the same." And he states this because an effect requires proper causes, and determinate effects require determinate causes, as [is said in] the second book of the *Physics*, and so a determinate demonstration requires a determinate principle.

The second conclusion: in demonstrating, one may not descend from genus to genus. Proof, from the text; since if one could do this, demonstration would not be from proper principles, but a descent could well be made from the genus that is indicated to another genus subalternate to it, just as to a species. For from mathematics, which is [composed] of two species, a single generic concept can be abstracted. And so geometry, which is one science, proceeds from one genus to another when it descends to the related genera of its many species, for example, isosceles, scalene, etc. So one should not interpret [the passage]

hic generis nullus secundum quamdam proportionem scientiis
ad scientiam, secundum illum modum loquendi quem facit Aristoteles infra,[33] quod una scientia est unius generis subiecti.

Tertia conclusio: nulla scientia procedit ex communibus. Patet
5   ex premissis, quia cause universalium universales sunt, et partic-
ularium particulares, ut in II *Physicorum*[34] et V *Methaphysice*,[35] et
ideo hic Aristoteles reprehendit[36] Brissonem, qui voluit probare
quadraturam circuli per hoc principium commune, ubicumque
est dare magis et minus, ibi est dare equale. Sed est dare quadra-
10   tum minorem circulo et maiorem; ergo equalem.

Et si dicas utrum ista propositio, 'maior assumpta, licet sit communis,' habet veritatem, dicunt aliqui quod in hiis que sunt eiusdem speciei, habet veritatem; et tu potes, si habes intellectum
eidem, hoc quomodo capiatur per Franciscum* in distinctione
15   prima, articulo primo,[37] et quomodo scientie subalternate utan-
tur communibus videbis infra.[38]

Quarta conclusio patet.

<Responsiones ad principalia>

Ad primum argumentum, patet responsio per primum notan-
20   dum, quia premisse reducuntur ad dignitates, vel etiam quia tractans de extremis, videtur tractare de mediis, et sic tractans de
subiecto et passione, tractat de premissis, que sunt media. Sed
prima mihi videtur utilior, sed magister Paulus,[39] qui impugnat
has responsiones verbaliter, impugnat.

25   Ad secundum, quod scientia methaphysicalis descendit in inferiores omnes ratione principiorum communium que tamen
sunt sibi propria. Unde hic probatur in textu, quod ipsa est maxime scientia, quia cum scire sit per causas, ipsa procedit per max-
imas causas, et ideo dicitur esse maxima scientia. Et hic utitur
30   Aristoteles[40] ista regula, sicut simpliciter ad simpliciter, ita magis
ad magis, et maxime ad maxime, et idem dico de dyalectica //
(fol. 19rb) cuius principiis alie scientie utuntur; de astrologia autem respectu naturalis similiter, quia

---

*Marg.*: "/ . . . / Franciscus format / . . . / maiorem sic ubi est / . . . / conocedens et ex
/ . . . / ibi equale / . . . / ipse negaret / . . . /ris clari excedens."

here in the mode of expression that Aristotle uses below, that one science is of one subject genus, to mean that nothing of the genus [descends] according to a kind of proportion from one science to another.

The third conclusion: no science proceeds from common principles. This is clear from the premises, since the causes of universal things are universal, and of particular things particular, as is said in *Physics* II and *Metaphysics* V. So at this place Aristotle criticizes Brisso, who wanted to prove the quadrature of the circle through the common principle that "Wherever there is more or less, there is something equal"; but it is possible to find a square smaller and larger than the circle; therefore [they are] equal.

And if you ask whether this proposition, "The major is assumed, granted it is common," is true, some say that when it refers to things that are of the same species, it is true; and you can [say this] if you make this assumption in the way assumed by Francis [of Mayronnes] in distinction 1, article 1. You will see below how subalternate sciences use common principles.

The fourth conclusion is obvious.

<*Responses to the Principal Arguments*>

To the first argument: the response is clear from the first notation, since premises are reduced to axioms, or also because in treating extreme terms one seems to appeal to middles, and thus in treating subject and property one treats of premises, which are [likewise] middles. The first seems to me to be more useful, but master Paul, who criticizes these responses verbally, argues against it.

To the second: metaphysical science descends into all inferior [sciences] by reason of common principles which nevertheless are proper to it. For this reason, the text here proves that metaphysics is science in the highest degree, since knowing scientifically is knowing through causes, [and] it proceeds through the highest causes, and so it is said to be science in the highest degree. And here Aristotle uses the following rule: just as absolute is to absolute, so the greater to the greater and the highest to the highest. I say the same of dialectic, whose principles are used by the other sciences. The same is true of astronomy with respect to natural philosophy, since

respectu subalternationis est hoc, quia quoad aliquam partem saltem, astrologia subalternatur naturali scientie, sed de dyalectica, certum est quod nullo modo subalternat sibi alias scientias. De methaphysica, autem est dicendum apud doctores quosdam,
5   ut infra videbitur.[41]

Ad tertium, patet responsio per secundam conclusionem.

Ad quartum, similiter.

Ad quintum, quod ymmo methaphysica procedit ex propriis, licet ista sint communia aliis pro magna parte.

10   Et ad confirmationem, dico quod etiam in generatione et putrefactione, licet sit virtus communis solis, tamen ista est contracta in aliqua materia respectu huius generis, et sic non est superfluum ponere principia communia, quia eorum virtutis principia inferunt effectum scientie.

15   Ad sextum, quod nec arismetica descendit in geometriam, nec econtra, tamen in terminis communibus invicem convertuntur, quia ratio numeri spectat ad arismeticam, et sic numerus cubus spectat ad eam,* licet cubo etiam utatur geometria, sed geometria bene descendit in perspectivam, et arismetica in armonicam, et
20   non tamen dicuntur descendere de genere in genus, quia se habent ut subalternans respectu subalternate.

Ad septimum, quod cirurgia subalternatur geometrie quoad istam conclusionem, et sic non est procedere ex extraneis, et quomodo per hoc, quod una scientia subalternat aliam scientiam,
25   subalternata non dicatur procedere ex principiis non propriis, Egidius[42] hic late protractat, inducendo exemplum de perspectivo, quod hanc passionem probat de oculo, quod de re visa videbitur oculum† destrum <esse destrum>, et sinistrum <esse sinistrum>. Et de hoc vide per eum hic, si volueris.

---

*eam ] eum C.
†oculum ] oculo C.

with respect to subalternation this is so, because astronomy is subalternated to natural science, at least in a certain part. But as far as dialectic is concerned, it is certain that it does not subalternate other sciences to itself in any way. As regards metaphysics, however, the [opinions of] some doctors must be discussed, as will be seen below.

To the third: the response is clear from the second conclusion.

To the fourth, the same.

To the fifth: even metaphysics proceeds from proper principles, although for the most part they are common to other [sciences].

To the confirmation, I say that even in [the case of] generation and putrefaction, although there is a common power in the sun, nevertheless it is contracted to a particular matter within the genus, and so it is not superfluous to assume common principles, since in virtue of them, the principles introduce the effect [proper to the] science.

To the sixth: arithmetic does not descend into geometry, nor vice versa; nevertheless in common terms they are convertible, since the formality number pertains to arithmetic, and so a cube number pertains to it, even though geometry also considers the cube. But geometry descends more readily into perspective and arithmetic into music, and yet they are not said to descend from one genus into another, since they are related as subalternating to subalternate.

To the seventh: [it must be said] that surgery is subalternated to geometry insofar as this conclusion is concerned, and so it does not proceed from extraneous principles, in this way, that [when] one science subalternates to itself another science, the subalternate [science] is not said to proceed from principles that are not proper. Giles deals extensively with this by adducing the example of the perspectivist, who proves this property of the eye, that with respect to the object seen, the right eye sees what is on the right and the left what is on the left. And regarding this, see what he says here if you wish.

Ad octavum, quod philosophia naturalis etiam subalternat sibi
medicinam, et de hoc infra dicetur, et vide hic Egidium,[43] qui respondit ad argumenta iam facta de medicina, et dyalectica, et
eclipsi, que fit per interpositionem terre, quia ad naturalem spec-
5    tat, vel in naturali vel in propria forma, quomodo stelle lumen
habeant, et si possint lumen perdere et qualiter, licet quomodo
per interpositionem terre fiat eclipsis, ut quia fit in capite vel in
cauda dragonis, spectat ad astrologiam.

Et etiam potest dari alia responsio* secundum ipsum, quod
10    astrologia est naturalis; modus autem considerandi est metha-
physicus, licet de hoc dicetur in questione speciali de subalternatione. Quare etc..

---

*Marg.: "Secunda responsio."

To the eighth: natural philosophy subalternates medicine to itself. This will be discussed below. See the response that Giles gives to the argument made here regarding medicine and dialectic and the eclipse that occurs by the interposition of the earth. Since the argument draws upon natural philosophy for the natural or the proper form, for the way in which the stars have light, and whether or in what way they can lose [their] light; but for the way in which the eclipse occurs through the interposition of the earth, [and] the fact that it takes place in the head or tail of the dragon, pertains to astronomy.

And according to [Giles] himself, another response can be given, that astronomy is itself natural [science], but its method of considering is metaphysical. This will be discussed in the special question on subalternation. Wherefore. . . .

## Notes

1. *Analytica Posteriora* I.7 75$^a$39–75$^b$2; *Aristoteles Latinus* IV.1–4, ed. L. Minio-Paluello and B. G. Dod (Bruges-Paris: De Brouwer 1968), 19.

2. *Analytica Posteriora* I.7 75$^b$13; *Aristoteles Latinus* IV.20.

3. *Analytica Posteriora* I.13 79$^a$15–16; *Aristoteles Latinus* IV.32.

4. *Analytica Posteriora* I.7 75$^a$38; *Aristoteles Latinus* IV.19.

5. *Analytica Posteriora* I.9 75$^b$37; *Aristoteles Latinus* IV.21.

6. *Analytica Posteriora* I.9 76$^a$26; *Aristoteles Latinus* IV.23.

7. *Analytica Posteriora* I.10 76$^a$31; *Aristoteles Latinus* IV.23.

8. Thomas Aquinas, *In libros Posteriorum Analyticorum Expositio* I, lect. 15, no. 131; ed. R. Spiazzi (Turin: Marietti 1964), 197–198.

9. *Analytica Posteriora* I.7 75$^a$38; *Aristoteles Latinus* IV.19.

10. *Analytica Posteriora* I.7 75$^b$14; *Aristoteles Latinus* IV.20.

11. Thomas Aquinas, *In libros Posteriorum Analyticorum Expositio* I, lect. 15, no. 133; *ed. cit.* p. 198.

12. *Analytica Posteriora* I.8 75$^b$24–25, 75$^b$31–33; *Aristoteles Latinus* IV.20–21.

13. *Physica* II.9 200$^a$34-$^b$7; *Aristoteles Latinus* VII.1, fasc. 2, ed. Fernand Bossier and Jozef Brams (Leiden-New York: E. J. Brill 1990), 94–95.

14. *Analytica Posteriora* II.3 90$^b$19ff.; *Aristoteles Latinus* IV.72ff.

15. *Forsitan Metaphysica* VII.15 1039$^b$28; *Aristoteles Latinus* XXV.2, ed. Gudrun Vuillemin-Diem (Leiden: Brill 1976), 151.

16. Thomas Aquinas, *In libros Posteriorum Analyticorum Expositio* I, lect. 16, no. 4–5; *ed. cit.* p. 200.

17. Cf. Antonius de Carlenis de Napoli, *Quaestiones in libros I–II Analyticorum Posteriorum Aristotelis* I, q. 16. conc. 3; fol. 18$^{ra}$.

18. *Analytica Posteriora* I.9 75$^b$37; *Aristoteles Latinus* IV.21.

19. *Analytica Posteriora* I.9 75$^b$40–76$^a$1; *Aristoteles Latinus* IV.21.

20. *Analytica Posteriora* I.9 76$^a$26; *Aristoteles Latinus* IV.23.

21. *Supra*, q. 17, (fol. 18$^{va}$), ed. p. 35.

22. Thomas Aquinas, *In libros Posteriorum Analyticorum Expositio* I, lect. 15, no. 130; *ed. cit.* p. 197.

23. Egidius Romanus, *Expositio in libros Posteriorum* (Venice: Bonetus Locatellus 1488), fol. D$_8$$^{ra}$.

24. Paulus Venetus, *Expositio in libros Posteriorum Aristotelis* (Venice: Simon Papiensis 1494), fol. 28$^{rb-va}$.

25. Paulus Venetus, *Expositio in libros Posteriorum Aristotelis; ed. cit.* fol. 28$^{vb}$.

26. Antonius de Carlenis de Napoli, *Quaestiones in libros I–II Analyticorum Posteriorum Aristotelis* I, q. 22; fol. 23$^{vb}$-25$^{rb}$ (edn. pp. 44–52).

27. *Analytica Posteriora* I.10 76$^a$37–76$^b$11; *Aristoteles Latinus* IV.23–24.

28. Ibid.

29. *Analytica Posteriora* I.2 71$^b$21; *Aristoteles Latinus* IV.7

30. *Analytica Posteriora* I.2 72$^a$6–7; *Aristoteles Latinus* IV.8.

31. *Physica* II.3 195$^b$26–27; *Aristoteles Latinus* VII.1, fasc. 2, p. 62.

32. *Analytica Posteriora* I.7 75ª38; *Aristoteles Latinus* IV.19.

33. *Analytica Posteriora* I.28 87ª38; *Aristoteles Latinus* IV.60.

34. *Physica* II.3 195ᵇ26–27; *Aristoteles Latinus* VII.1, fasc. 2, p. 62.

35. *Metaphysica* V.2 1013ᵇ34; *Aristoteles Latinus* XXV.2, p. 86.

36. *Analytica Posteriora* I.9 75ᵇ40–76ª1; *Aristoteles Latinus* IV.21. *De sophisticis elenchis* XI 171ᵇ16, 172ª4; *Aristoteles Latinus* VI, ed. B. G. Dod (Leiden-Bruxelles: Brill, De Brouwer 1975), 25, 26.

37. Franciscus de Mayronis, *In IV libros Sententiarum*, I, d. 1, q. 1, a. 1 (Venice: Octavianus Scotus 1520), fol. 12ᵛᵃ⁻ᵇ M-N.

38. Antonius de Carlenis de Napoli, *Quaestiones in libros I–II Analyticorum Posteriorum Aristotelis* I, q. 22; fol. 23ᵛᵇ-25ʳᵇ, esp. 24ʳᵇ (edn. p. 46).

39. Paulus Venetus, *Expositio in libros Posteriorum Aristotelis; ed. cit.* fol. 28ʳᵇ⁻ᵛᵃ.

40. *Metaphysica* I.2 982ª22–982ᵇ7; *Aristoteles Latinus* XXV.2, pp. 9–10.

41. Antonius de Carlenis de Napoli, *Quaestiones in libros I–II Analyticorum Posteriorum Aristotelis*, I, q. 22; fol. 23ᵛᵇ-25ʳᵇ, esp. 24ᵛᵇ (edn. p. 48).

42. Egidius Romanus, *Expositio in libros Posteriorum* (Venice: Bonetus Locatellus 1488), fol. E₂ᵛᵇ.

43. Egidius Romanus, *Expositio in libros Posteriorum; ed. cit.* fol. D₈ᵛᵃ.

## &lt;L. I, Q. 22: 'UTRUM CONVENIENTER ASSIGNETUR AB ARISTOTELE DISTINCTIO SCIENTIE SUBALTERNANTIS ET SUBALTERNATE'*&gt;

Vicesimo secundo, queritur utrum conventienter assignetur ab
5   Aristotele distinctio scientie subalternantis et subalternate. Et arguitur quod non, quia sic penes subiectum contractum et non
contractum, // (fol. 24ra) et non contractum per aliquam differentiam, fieret subalternatio, et sic sequeretur quod methaphysica
subalternaret omnes scientias sibi, quod tamen est falsum, cum
10   multe scientie sint que habent principia nota per se, que propter
hoc, cum non possint probari per scientiam methaphysicalem,
non poterunt sibi subalternare. Ergo penes subiectum contractum et non contractum, non fit subalternatio, et secundum
Commentatorem,[1] quia scientia subalternans debet probare prin-
15   cipia scientie subalternate, et tamen in philosophia naturali habemus principia que sunt per se notissima, ut naturam esse et
motum esse.

Secundo, si hec haberet veritatem, sequeretur quod geometria
subalternaretur philosophie naturali; sed hoc est falsum. Et pro-
20   batur consequentia, quia in philosophia naturali etiam principia
geometrica probantur, sicut est illud de punto ad puntum contingit rectam lineam ducere, ut probatur VI *Physicorum*[2]; et tamen
hic dicit Aristoteles quod propter probationem principiorum,
una scientia alteri subalternatur.

25   Tertio, ex hoc sequitur quod liber *De anima* subalternaretur libro
*Physicorum*, et haberet subiectum magis contractum,[†] et etiam
quia assumit principia libri *Physicorum*.

Quarto, unus liber philosophie subalternaretur alteri, quia accipit una scientia principia alterius.

30   Quinto, ex dictis Aristotelis,[3] sequitur quod perspectiva et arismetica subalternantur philosophie naturali. Probatur consequentia, quia capiendo scientiam perspective, manifestum est quod
subalternatur geometrie et armonica subalternatur arismetice, et

---

*C fol. 23vb.
[†]*Marg.:* "Nota."

44

# &lt;BOOK I, QUESTION 22: WHETHER THE DISTINCTION BETWEEN THE SUBALTERNATING AND SUBALTERNATE SCIENCES IS SPECIFIED APPROPRIATELY BY ARISTOTLE&gt;

In the twenty-second, the question is whether the distinction between subalternating and subalternate sciences is properly specified by Aristotle. The argument is that it is not, since subalternation occurs between subjects that are contracted and those that are not contracted by any differentia; from this it should follow that metaphysics subalternates all the other sciences to itself; but this conclusion is false, since there are many sciences that have principles that are self evident, and on this account, since these principles cannot be proved through metaphysical science, they cannot be subalternated to it. Therefore, subalternation does not occur between subjects that are contracted and those that are not. According to the Commentator, the subalternating science must prove the principles of the subalternate science, and yet in natural philosophy we have principles that are most self evident, as for example, that nature exists and motion exists.

Second, if this were true, it would follow that geometry would be subalternated to natural philosophy; but this is false. Proof of the consequence: geometrical principles are proven in natural philosophy, as for example, that it is possible to draw a straight line between two points, as is proved in *Physics* VI; and yet Aristotle says here that one science is subalternated to another through the proof of its principles.

Third, it follows from this that *De anima* would be subalternated to the *Physics*, and it would have a subject that was more contracted, and it would also take its principles from the *Physics*.

Fourth, one book of philosophy would be subalternated to another because one science takes its principles from another.

Fifth, from Aristotle's statements it follows that perspective and arithmetic are subalternated to natural philosophy. Proof of the consequence: taking the science of perspective, it is clear that it is subalternated to geometry, and music is subalternated to arithmetic, and

tamen quelibet istarum subalternatur philosophie naturali, cum subiectum sit contractum ad materiam sensibilem.

Sexto, dyalectica est scientia communis quantum ad subiectum, et etiam quantum ad principia eius quibus omnes alie scientie utuntur. Ergo cum hoc sit falsum, sequitur quod male dicat Aristoteles.

Septimo, scientia de yride, que habetur in libro *Meutarorum*,[4] accipit quedam principia perspective, cum perspectiva sit de linea visuali sine radiosa, et tamen sibi non subalternatur.

Octavo, in loyca, scientia sex principiorum accipit principia posita ab Aristotele in presentis, et tamen sibi non subalternatur.

Nono, ex hiis que dicit Aristoteles in textu, sequitur quod quelibet scientia subalternata subalternaretur duabus scientiis, sicut adducit Landulfus in I *Sententiarum*,[5] et dat exemplum de perspectiva quoad hoc principium, 'Linea visualis recta est perfectissima,' quia una pars istius probatur in scientia naturali per illud principium, 'Agens naturale aproximatum passo perfectius agit,' et etiam probatur per scientiam geometrie. Et sic scientia perspective subalternatur geometrie et etiam scientie naturali.

Decimo, accipiamus theologiam; ipsa habet subiectum magis contractum quam methaphysica, et tamen sibi non subalternatur, ut manifestum est.

Undecimo, aut ista subalternatio que sit per illud subiectum contractum sit quia est de aliqua specie superioris—et cum genus et species reducantur ad eandem scientiam, sequitur quod scientia subalternans et subalternata erunt simpliciter eadem scientia—, aut erit contractum non ut species, sed per aliquod additum accidentale. Et tunc sequitur // (fol. 24rb) quod illa scientia subalternata semper erit de ente per accidens, quod est contra Aristotelem supra[6] que est de eodem libro.

Duodecimo, secundum predicta, stereometria est subalternata geometrie, et tamen non est verum, quia stereometria est eadem scientia cum geometria, cum sit de mensuris corporum sicut et accidentia.

Tertio decimo, astrologia secundum predicta esset subalternata geometrie, et similiter scientia navalis, que dicitur apparentie, et tamen iste dicuntur esse scientie naturales. Ergo male Aristoteles distinxit in textu isto.

yet each of these [sciences] is subalternated to natural philosophy, since its subject is contracted to sensible matter.

Sixth, dialectic is a common science with regard to its subject, and even with regard to its principles, which all the other sciences use. Therefore, since this is false, it follows that Aristotle spoke improperly.

Seventh, the science of the rainbow, which is treated in the *Meteorologies*, takes some of its principles from perspective, since perspective concerns visual lines without their being radiant, and nevertheless it is not subalternated to perspective.

Eighth, in logic, the science of the six principles takes principles posited by Aristotle in the *Posterior Analytics*, and yet it is not subalternated to the *Analytics*.

Ninth, from what Aristotle says in the text, it follows that every subalternate science should be subalternated to two sciences, just as Landulf adduces in the first book of the *Sentences*. He gives the example of perspective as it concerns this principle, "A straight visual line is the most perfect," since one part of the [principle] is proven in natural science through the principle, "The closer the natural agent is to the patient, the more perfectly it acts," and the same is proved through the science of geometry. So the science of perspective is subalternated to geometry and also to natural science.

Tenth, let us take theology. It has a subject more contracted than has metaphysics, and yet it is not subalternated to metaphysics, as is clear.

Eleventh, either this subalternation occurs through the contracted subject because it is some species of the superior [science]—and, since the genus and the species are reducible to the same science, it follows that the subalternating and subalternate science will be strictly speaking the same science—, or it will be contracted not as a species but through some accidental addition. Then it follows that the subalternate science will always concern accidental being, which is contrary to what Aristotle says above in the same book.

Twelfth, according to what has been said above, stereometry is subalternated to geometry, and yet this is not true, since stereometry is the same science as geometry, being concerned with the measure of bodies as with their other accidents.

Thirteenth, astronomy, according to what has been said, should be subalternated to geometry, and similarly navigation, which is called [the science] of appearances, and nevertheless the latter are said to be natural sciences. Therefore Aristotle distinguished improperly [between the subalternating and subalternate sciences] in this text.

Quarto decimo, quia subalternans et subalternata univocantur, ut dicitur hic in textu[7]; et sic sequitur quod nulla scientia alterius speciei subalternaretur alteri, quod est expresse falsum.

In contrarium,* est textus hic *alio autem modo differt*,[8] ubi assig-
5  natur differentia scientie subalternantis et subalternate penes principia contracta et penes subiecta.

### \<Alique notanda\>

Primo, notandum quod ut notat hic sanctus doctor,[9] illa scientia dicitur subalternare que est dicens propter quid, et sic una sci-
10  entia est dupliciter subalternata. "Uno modo, quando subiectum unius scientie est species superioris scientie, sicut animal est species corporis naturalis, et ita scientia de animalibus est sub sci-
entia naturali.

Alio modo, quando subiectum inferioris scientie non est spe-
15  cies subiecti superioris, sed comparatur ad subiectum superioris, sicut materiale ad formale. Et hoc modo, accidit esse una scientia sub altera, ut perspectiva se habet ad geometriam." Unde "linea visualis non est species linee simpliciter, sicut nec triangulus lig-
neus est species trianguli simpliciter, et similiter machinativa, id-
20  est scientia de machinis, se habet ad stereometriam, idest ad scientiam que est de commensurationibus corporum," et idem de musica respectu arismetice. Et sic postea[10] sanctus doctor declarat quomodo principia mathematicalia applicantur ad sensi-
bilia in perspectiva et musica. Et sic aliquis potest scire propter
25  quid, nesciens quia; sic etiam concludit quod scientia de yride subalternatur perspective, quia applicat principia que perspectiva tradit ad materiam determinatam, et similiter de cirurgia, ubi dicit Aristoteles[11] quod aliquando \<propter\>[†] quid et quia differunt in scientiis non subalternantibus, et sic ad aliquam conclu-
30  sionem in medicina consideratam, applicabilia sunt principia geometrie, sicut quod 'vulnera circularia tardius sanantur.' Med-
ici est scire quia per hoc quod[‡] experitur, sed propter quid est geometrice ad quam partium cognitio scire, quod si circulus est figura sine angulo, unde partes circulares vulneris non appropin-
35  quant sibi invicem ut posset coniungi de facili.

---

*Marg.:* "In oppositum."
[†]propter *om.* C.
[‡]quod *add. supra lin.* C.

Fourteenth, because subalternating and subalternate are univocal, as is said here in the text; and so it follows that no science is subalternated to another [science] of a different species, which is expressly false.

To the contrary is the text here "However, in another way, it differs . . . ," where the differentia of the subalternating and subalternate sciences is assigned according to principles that are contracted principles and according to subjects.

<Some Notations>

First, note that as St. Thomas observes here, a science is said to subalternate when it gives the reason why [*propter quid*], and so one science is subalternated in two ways. "In one way, when the subject of one science is a species of the superior science, as animal is a species of natural body, and so the science of animals is under natural science.

In another way, when the subject of the inferior science is not a species of the superior subject, but is compared to the subject of the superior science as material to formal. And in this way, it happens that one science falls under the other, as perspective is related to geometry." Whence "a visual line is not a species of line absolutely, just as a wooden triangle is not a species of triangle absolutely, and in the same way mechanics, that is, the science of devices, is related to stereometry, that is, the science which deals with the measurement of bodies," and the same is true of music with respect to arithmetic. And so later St. Thomas specifies how mathematical principles are applied to sensible things in perspective and music. Thus a person can know the reasoned fact [*propter quid*] without knowing the fact [*quia*]. So he also concludes that the science of the rainbow is subalternated to perspective, since it applies principles treated in perspective to determinate matter. The same applies to surgery, where Aristotle says that sometimes the reasoned fact [*propter quid*] and the fact [*quia*] are distinguished in sciences that are not subalternating. So the principles of geometry are applicable to some conclusions considered in medicine, as for example, that "circular wounds heal more slowly." It is for the physician to know the fact that this occurs, but the reason for the fact pertains to the geometer, who has knowledge of the parts, so that if a circle is a figure without angles, the parts of a circular wound are not close to each other in such a way that they can be joined together easily.

Secundo, notandum quod ut dicit hic textus,[12] pure mathematice sunt de subiecto magis abstracto et secundum speciem,
sed scientia subalternata ut perspectiva et arismetica sunt de subiecto contracto, et dicuntur esse de subiecto secundum materiam,
5    ut ait Philosophus in II *Physicorum*,[13] iste scientie inferiores dicuntur esse mathematice, et sanctus doctor expresse dicit[14] quod
magis sunt dicende physice quam mathematice, quia subiecta
earum sunt contracta ad materiam sensibilem. Et sic etiam dicit
Burleus[15] // (fol. 24va) exponens Commentatorem, sed Egidius[16]
10   ponit hic ut ponit sanctus doctor, et postea relinquit sub dubio.

Et hoc attendendum* quod secundum doctrinam sancti doctoris infra in hoc libro,[17] ubi loquitur de unitate scientie ex distinctione eiusdem, potest oriri dubium ex eo, quod ipse semper
habet pro regula generali quod distinctio scientiarum sit penes
15   media, cum enim ipse dicat quod media harum scientiarum sub-
alternatarum, scilicet perspective et arismetice, sint mathematice, sequitur quod hec due scientie erunt magis mathematice
quam naturales, cuius oppositum noverit ipse ubi supra.

Sed potest responderi† quod etiam media harum scientiarum
20   proximam connotant materiam intellectu insensibilem, sicut
etiam subiecta earum, et sic magis debent dici naturales quam
mathematice.

Et hic etiam advertis‡ quia circa hanc conclusionem positam,
notat Egidius,[18] sicut et sanctus doctor,[19] quod mathematicus ab-
25   strahit a materia quali; non tamen abtrahit a materia simpliciter,
quia non a materia quanta, quia plus in materia est quantitas
quam qualitas, et omne prius potest intelligi sine suo posteriori,
et sic potest intelligi in aliquo esse quantitatem, non intelligendo
esse qualitatem. Sed magister Paulus[20] hic dicit§ contradicens,
30   'Ego autem dico# quod licet mathematicus non abstrahat a materia intelligibili, que est quantitas continua vel discreta,** abstrahit
tamen a materia intelligibili, que est materia prima considerata
absque forma, et a materia sensibili, que est eadem materia considerata cum quantitate sensibili forma, quia licet quantitas sit et
35   prius insit substantie secundum

---

*Marg.: "Dubium."
†Marg.: "Solutio."
‡Marg.: "Nota."
§Marg.: "Opinio magistri Pauli contra Egidium."
#Marg.: "Nota quod mathematicus non abstrahit a materia que est substantia, neque a
materia que est quantitas."
**abstrahit tamen a materia intelligibili que est quantitas continua vel discreta *add. sed
canc.* C.

Second, note that as the text says here, pure mathematics treat a subject that is more abstract and specific, but subalternate sciences like perspective and arithmetic treat a contracted subject, and are said to treat a subject according to its matter. As the Philosopher says in *Physics* II, these inferior sciences are said to be mathematical, and St. Thomas expressly states that they are called more physical than mathematical, since their subjects are contracted to sensible matter. And even Burley says the same thing while interpreting the Commentator, but Giles establishes the point here in the way St. Thomas does, though later he calls it into doubt.

Observe that according to the teaching of St. Thomas later in this book, where he is speaking of the unity of science in conjunction with their distinction, a doubt can arise because he himself always takes as a general rule that the distinction of the sciences should be taken from the middle terms, for when he says that the middle terms of these subalternate sciences, namely, perspective and arithmetic, are mathematical, it follows that these two sciences will be more mathematical than natural, a view contrary to that expressed above.

But one can reply that the middle terms of these sciences connote a proximate matter that is imperceptible to the intellect, as are their subjects, and so they can be said to be more natural than mathematical.

Here also note that concerning this conclusion, Giles observes just as does St. Thomas that the mathematician abstracts from matter of a particular kind but not from matter absolutely. For he does not abstract from matter as quantified, because quantity is in matter prior to quality, and everything that is prior can be understood without the posterior; thus quantity can be understood to exist in a subject, even though quality does not. But master Paul, disputing this, says, "But I say that although the mathematician does not abstract from intelligible matter, understood as continuous or discrete quantity, he does abstract from intelligible matter understood as prime matter considered without form and from sensible matter which is the same matter considered with quantitative, sensible form. For, although (according to the Commentator in *Metaphysics* XII) quantity exists and inheres primarily in substance

Commentatorem[21] XII *Methaphysice,* et communiter presupponit substantiam ordine nature, quia tamen non includit substantiam in suo conceptu quidditativo, potest mathematicus abstrahere quantitatem ab omni substantia. Et ideo dicit Philosophus[22] quod

5　licet geometralia insint subiecto, tamen intelliguntur sine subiecto, quia non circumcernunt subiectum, ac si essent sine subiecto.'

Sed contra hoc tantum magistri Pauli, arguitur* quia subiecta cadunt in diffinitione suorum accidentium, quod probatur quia

10　sicut se habet res ad esse, ita se habet ad cognosci, ut in II *Methaphysice,*[23] et que sunt causa in essendo sunt causa in cognoscendo, ut in VI *Thopicorum,*[24] et quodlibet accidens est entis, ut in VII *Methaphysice.*[25] Ergo in diffinitione accidentis quod est quantitas, si quidditative diffiniatur, poneretur subiectum etiam;

15　de hoc dictum est supra in questione.[26] Sed ne videatur concludi ex hiis dictis quod geometra debeat cognoscere conceptus substantie dum vult diffinire conceptus quantitatis, dico quod geometra non diffinit quidditatem quantitatis, sed diffinit proportiones sive commensurationes et talia sub quibus conveniat de

20　quantitate.

Tertio, notandum quod sicut patebit in responsionibus ad argumenta, est dubium si ex hoc, quod subiectum contineatur sub subiecto, sequatur subalternari illi scientie. Et manifeste patet quod si hoc esset verum, omnis scientia subalternaretur metha-

25　physice, que est de ente simpliciter secundum Commentatorem,[27] vel si vis dicere secundum genera est de ente ut est reducibile in primam formam et ultimum finem. Et ideo Egidius in I *Physicorum,*[28] pertractans de // (fol. 24vb) scientia naturali si subalternatur methaphysice, dicit quod non, quia in scientia na-

30　turali sunt principia per se nota; et ex hoc etiam improbat ibi dictam sancti doctoris,[29] qui posuit quod ens mobile est subiectum, quia sic videtur contrahere ens ad aliquem modum entis, et sic subalternatur methaphysice. Sed de hoc est alia questio; ideo ad presens dimitto, quia idem est quod sanctus doctor[30] fundat se

35　super aliis rationibus, ponens ens mobile esse subiectum in philosophia naturali, et ita tenet. Et licet hic in ista ratione, videatur dicere quod sit substantia naturalis, et supra de erroribus universalium, quod sit corpus mobile. Sed hoc dixi loquendo exemplariter.†

---

*Marg.:* "Impugnatio magistri Pauli."

†*Marg.:* "Nota quod sanctus doctor hic videtur dicere quod substantia(?) naturalis est subiectum in philosophia et supra de erroribus / . . . / universale dicere quod corpus mobile, sed locutus est exemplariter."

and commonly presupposes substance in the order of nature, since it does not include substance in its essential definition, the mathematician can abstract quantity from any substance. Thus the Philosopher says that although geometricals inhere in a subject, nevertheless they are understood without a subject, since they do not presuppose a subject, as if they were to exist without a subject."

But against this position only master Paul argues, saying that subjects are contained in the definition of their accidents. His proof: because just as things are related in being so they are related in knowing, as in *Metaphysics* II, and what are causes in being are also causes in knowing, as in *Topics* VI, and every accident is [an accident] of a being, as in *Metaphysics* VII. Therefore in the definition of the accident that is quantity, if it is defined essentially, a subject should also be included; this has been discussed in a question above. But lest one gather from these remarks that the geometer must know the concept of substance when he wishes to define the concept of quantity, I say that the geometer does not define the essence of quantity, but he defines ratios or measurements and such things among those that belong to quantity.

Third, note that just as will be clear in the responses to the objections, there is some doubt whether sciences are subalternated simply from the fact that one subject is contained under another subject. It is manifestly clear that if this were true, every science would be subalternated to metaphysics, which is concerned with being simply, according to the Commentator, or if you wish to say generically that it is concerned with being as reducible to first form and final end. Therefore in discussing whether natural science is subalternated to metaphysics in *Physics* I, Giles says that it is not, since in natural science there are principles that are self-evident. From this he refutes the position of St. Thomas, who held that changeable being is its subject, since in this way it seems to contract being to some type of being, and so [natural science] would be subalternated to metaphysics. But this is another question, and so at present I refrain from discussing it, since St. Thomas established this conclusion by other arguments, positing that changeable being is the subject in natural philosophy, and this is what he holds. And although here in this argument it appears that its subject is natural substance, and above in the context of the errors concerning universals, that it is the changeable body, I have said that this was said by way of example.

Quarto, notandum quod ad subalternationem proprie dictam
requiruntur predicta duo, videlicet subiectum esse sub subiecto,
ita quod subiectum scientie subalternantis sit contractum per aliquam differentiam accidentalem, et ulterius quod scientia supe-
5  rior dicat propter quid de eodem de quo scientia inferior dicit
quia. Si autem ista duo non concurrant, non dicitur proprie subalternatio. Et sic patet de questione illa que est famosa in theologia, de qua sanctus doctor cum sequentibus eum, quod
subalternatur scientie beatorum[31]; Scotus autem cum sequenti-
10 bus eum tenet quod nullomodo.[32] Sed Egidius,[33] sequens sanctum doctorem ex hoc textu, colligit tres conclusiones scientie
subalternate et subalternantis que competunt illi scientie. Sed de
hoc diximus in I *Sententiarum*.[34]

### \<Quattuor conclusiones\>

15 Prima conclusio: scientia subalternans semper est certiora
quam scientia subalternata. Probatur quia scientia de magis abstracto et dicens causam est certiora secundum se quam scientia
que est de minus abstracto, et que dicit quia tantum. Et ad hanc
conclusionem faciunt omnes dicere posite ab Aristotele in textu,[35]
20 videlicet quod scientia subalternans et subalternata differunt in
modo sciendi, et quod differunt in modo rei scite, et quod differant in modo generandi, que omnia reducuntur ad predictam
conclusionem.

Secunda conclusio: nec methaphysica subalternat sibi alias sci-
25 entias, nec philosophia* ceteras partes philosophie. Illa conclusio
quantum ad utramque partem probatur ex dictis supra per Egidium in secundo notando,[36] quia physica habet principia per se
nota, et secunda pars probatur ex dictis sancti doctoris hic in isto
textu, quia contractio subiecti facta in ceteris partibus philosophie
30 non sufficit ad hoc, quod ille partes subalternentur libro *Physico-*
*rum*, et de hoc dictum est in primo notando per sanctum doctorem.[37]

---

*add. in marg.:* "alias scientias, nec philosophia."

Fourth, note that the aforesaid two things are required for sub-alternation properly speaking, namely that one subject be contained under the other, with the subject of the subalternating science being contracted by an accidental differentia, and beyond this that the superior science give the reasoned fact for what the inferior science gives the fact. If these two conditions are not met, it is not properly said to be subalternation. And so, concerning the question famous in theology, argued by St. Thomas and those following him, it is clear that theology is subalternated to the science of the blessed; however, Scotus and those following him hold that this is not the case. But Giles, following St. Thomas on this text, gathers three conclusions about the subalternate and subalternating sciences that are proper to that science. We have spoken about this in the first book of the *Sentences*.

<*Four Conclusions*>

The first conclusion: a subalternating science is always more certain than a subalternate science. Proof: a science that concerns the more abstract and speaks to the cause [of a fact] is more certain in itself than a science that concerns the less abstract and speaks only to the fact. Everyone comes to this conclusion, taken from Aristotle in the text, namely, that subalternating and subalternate sciences differ in their way of knowing, in the mode of the thing known, and in the mode of generation, all of which are reduced to the aforesaid conclusion.

The second conclusion: metaphysics does not subalternate other sciences to itself, nor does [first] philosophy subalternate other parts of philosophy. Both parts of this conclusion are proved from what has been said above by Giles in the second notation, since physics has principles that are self-evident. The second part [of the conclusion] is proved from the words of St. Thomas here in this text, since the contraction of the subject made in other parts of philosophy does not suffice for those parts to be subalternated to the *Physics*. This has already been discussed from the viewpoint of St. Thomas in the first notation.

Tertia conclusio: non inconvenit eandem esse scientiam subalternantem et subalternatam secundum diversas partes. Probatur specialiter de philosophia naturali, quia quantum ad illam
partem que est de yride, subalternatur perspective, et tamen
5 perspectiva secundum aliud potest dici subalternari philosophie naturali. Et hic attendendum quod Aristoteles in fine huius
textus, ubi dicit *multe enim et non sub invicem scientiarum*[38] etc.,
quia secundum Egidium[39] et magistrum Paulum,[40] cirurgia,
que* dicit quod vulnera circularia tardius sanantur, non subalter-
10 natur geometrie, tamen concedunt quod quoad istam partem
subalternatur.
Quarta conclusio patet.

### <Responsiones ad argumenta principalia>

Ad primum, patet responsio per secundum notandum, quia
15 quando scientie habent principia per se nota, non subalternantur
alteri, licet subiectum sit sub subiecto. // (fol. 25ra) Dicit metha-
physicus respectu scientie naturalis, et ad hoc potest induci quod
Aristoteles, volens procedere contra Parmenidem et Mellissum,
subinduit habitum methaphysici. Unde dicit quod habet hic
20 prime philosophie respectus, ut in I *Physicorum*.[41]
Ad secundum, quod non inconvenit respectu alicuius principii
subalternari geometriam philosophie naturali, sic in exemplo
predicto arguendo. Et sic confirmatur tertia conclusio, quia scientia naturalis, quantum ad scientiam de yride, subalternatur
25 perspective, que subalternatur geometrie, et tamen geometria,
quantum ad predictum principium, subalternatur philosophie,
licet ut dictum est ibidem, non dicatur esse subalternatio, ut
demonstratum est.
Ad tertium, patet responsio per secundam conclusionem, et
30 per dicta sancti doctoris in textu.
Ad quartum, quod liber *Physicorum* presupponit principia sua;
sed bene utitur quibusdam principiis methaphysicalibus, ut ip-
semet exponit in I *Physicorum*,[42] ut declaratum est.

---

*que *add. supra lin.* C.

The third conclusion: nothing prevents the same science from being subalternating and subalternate under different formalities. The proof, especially with regard to natural philosophy: with respect to the part that concerns the rainbow, it is subalternated to perspective, and yet on other topics perspective can be said to be subalternated to natural philosophy. Here note what Aristotle says at the end of this text, where he says "for many of the sciences which are not under one another . . . ," since according to Giles and master Paul, surgery, in showing that circular wounds heal more slowly, is not subalternated to geometry, and still they concede that at least in this matter it is subalternated.

The fourth conclusion is clear.

<Responses to the Principal Arguments>

To the first [argument]: the response is clear from the second notation, because when sciences have principles that are self-evident, they are not subalternated to another science, even though one subject may be under another. The metaphysician is said with respect to natural science, and to this one might add that in his desire to proceed against Parmenides and Melissus, Aristotle took on the role of the metaphysician. Whence he speaks here with respect to first philosophy, as in *Physics* I.

To the second, there is nothing wrong with geometry being subalternate to natural philosophy with respect to a particular principle, as was argued in the example discussed above. And so the third conclusion is confirmed, because natural science with regard to the science of the rainbow is subalternated to perspective, which is subalternated to geometry, and nevertheless geometry with respect to the aforesaid principle is subalternated to [natural] philosophy, although as has been said in the same place, this should not be called subalternation, as has been shown.

To the third, the response is clear from the second conclusion and by what St. Thomas says in the text.

To the fourth, the *Physics* presupposes its own principles, but it uses some metaphysical principles well, as Aristotle himself explains in *Physics* I, as has been stated.

Ad quintum, quod ille due scientie, scilicet perspectiva et armonica vel musica, non subalternantur philosophie naturali, nec
Aristoteles hoc aliquando dixit; unde perspectivus generat de
linea visuali, et sic visuale contrahit lineam. Et sic de musica re-
5   spectu arismetice quantum ad numerum.

Ad sextum, quod dyalectica dicitur scientia communis, quia ea
alie scientie utuntur in discurrendo vel in arguendo, et non quod
ipsa probet principia aliarum.

Ad septimum, quod ymmo scientia de yride subalternatur
10  perspective.

Ad octavum, dicitur quod *Liber sex principiorum* non subalternatur libro presentorum, sicut nec pars subalternatur alteri parti,
quia Gilibertus Porretanus ad declarandum partes presentes fecit
illum librum.

15  Ad nonum, quod illa propositio, quod scientia subalternata
subalternatur duabus scientiis, posita* a Landulfo tanquam
necessaria[43]; sed hoc non reperio in textu Aristotelis, nec in aliquo alio doctore. Et si esset aliquod bonum argumentum, adhuc
esset istud quod dictum est in hoc argumento. Et si diceretur
20  quod perspectiva subalternatur geometrie, et quia non reperitur
< . . . >,† non oportet hoc simpliciter affirmari; sed quantum ad
propositum ad quod < . . . >,‡ satis responsum est ibidem.

Ad decimum, quod principia theologie nullo modo declarantur
in methaphysica, cum illa sint revelata, ut omnes concedunt.

25  Ad undecimum, patet responsio per primum notandum, quia
species alicuius generis non causat subalternationem. Et ita argumentum illud est difficile, quia videtur sequi secundum sancti
doctoris dicta[44] in secundo notando, sicut etiam alii doctores sequuntur eum, quod perspectiva, cum sit de linea visuali, est de
30  ente per accidens.§ Sed ad hoc potest responderi quod linea visualis est circumlocutio unius generis subicibilis in illa scientia,
et similiter de numero sonoro respectu musice, licet sint quedam
alia contra talem responsionem,# quia de ente per accidens ut de
subiecto, potest formari aliqua propositio per se, de qua potest
35  esse scientia, et hec responsio dicta est supra in questione < . . .
>**.[45] Et licet sit ibi ad propositum, tamen hic non esset ad pro-
positum, quia queritur de ipso per se subiecto huius scientie;
nam de eo cui competit esse per accidens, non est scientia, sicut
plene declaratum est in // (fol. 25rb) predicta questione. Cum ergo
40  linea visualis sit tale aggregatum cui competit esse per accidens,
de eo non erit scientia.

---

*posita ] positis C.
†< . . . > *lacuna* C.
‡< . . . > *lacuna* C.
§*Marg.:* "Nota quomodo scientia subalternata est de subiecto per accidens, et quomodo non."
#*Marg.:* "Nota quod Antonius Andreas in tribus principiis concedit hoc de scientia subalternata."
**< . . . > *lacuna* C.

To the fifth: these two sciences, namely perspective and harmonics or music, are not subalternated to natural philosophy, nor did Aristotle say this in any place; whence the perspectivist generates the visual line, and so 'visual' contracts 'line'. The same [is true] of music with respect to arithmetic insofar as numbering is concerned.

To the sixth, dialectic is called a common science because other sciences use it in disputing or arguing, not because it proves the principles of other [sciences].

To the seventh: the science of the rainbow is itself subalternated to perspective.

To the eighth: the *Book of six principles* is not said to be subalternated to the *Posterior Analytics*, nor is one part of it subalternated to another, because Gilbert de la Porrée used that book to explain parts of the *Analytics*.

To the ninth: the proposition that a subalternate science is subalternated to two sciences was proposed by Landulf as necessary, but I do not find this in Aristotle's text nor in any other doctor. And if there were some good argument [for this position], what has been said in this argument would still obtain. And if it should be said that perspective is subalternated to geometry, and because it is not found < . . . > it should not be held absolutely; but with regard to what has been proposed < . . . >, the reply there is sufficient.

To the tenth: the principles of theology are in no way found in metaphysics, since they have been revealed, as all concede.

To the eleventh: the response is clear from the first notation, since being a species within a genus does not cause subalternation. Yet that argument offers a difficulty, since it seems to follow from the remarks of St. Thomas in the second notation and the other doctors who follow him, that because it concerns the visual line, perspective treats of being *per accidens*. But to this one can reply that 'visual line' is a circumlocution for one genus subject in that science, and that 'sounding number' with respect to music is another. Yet there are other arguments against this response, because with respect to a being *per accidens* as a subject, a proposition can be formed that is *per se*, and of this there can be science. This response has been made above in question < . . . >. Although it may be pertinent there, it is not here, because what is sought is the *per se* subject of this science. There is no science of a being *per accidens*, as was plainly stated in the aforesaid question. Since therefore visual line is the kind of aggregate that has being *per accidens*, there will be no science regarding it.

Ad duodecimum, quod ut respondit Egidius,[46] aliqui dixerunt quod stereometria, de qua loquitur hic Aristoteles,[47] <est subalternata>* geometrie in quo ageri de solidis et de corporibus. Possumus tamen dicere quod stereometria, que dicit propter

5  quid eius de quo machinativa dicit quia, est subalternata geometrie, quia geometria est abstracta magis, sed machinativa est magis materialis, stereometria est quedam practica geometrie.

Ad tertium decimum, quod astrologia est scientia naturalis, et principia sui indigent geometria; navalis nota, sive apparentia,

10  est generans per quedam signa apparentia, ut probatur quia luna facit circulum vel quia rubet celum, sed abstrologia cognoscat modo magis suttili. Unde dicitur esse de eo quod est medium inter necessarium et possibile.

Ad quartum decimum, dicitur quod quedam fere sunt univoce,

15  quia communiunt in nomine generis subalternans et subalternata; non autem in nomine speciei, licet differenter hoc contingat, secundum quod magis et minus appropinquant ad invicem.

---

*<est subalternata> *conj.; lacuna* C.

To the twelfth, as Giles responds, some say that the stereometry of which Aristotle speaks here [is subalternated] to geometry in that it deals with solids and bodies. Nevertheless, we can say that stereometry, insofar as it gives the reasoned fact of matters for which mechanics gives the fact, is subalternated to geometry, since geometry is the more abstract whereas mechanics is the more material, [and] stereometry is a kind of practical geometry.

To the thirteenth: astronomy is a natural science, and its principles depend on geometry; the science of navigation or stargazing proceeds through certain visible signs, when proving why the moon makes a circle or why the sky becomes red, but astronomy understands this in a more profound way. Whence it is said to concern that which is intermediate between the necessary and the possible.

To the fourteenth: some [sciences] are nearly univocal, since the subalternating and subalternate sciences agree in a generic name; however, they do not [agree] in the specific name, although this can occur in various ways according as they are nearer or farther from each other.

## Notes

1. *Aristotelis Posteriorum Resolutoriorum libri duo cum Averrois . . . magnis commentariis. . . .* I, comm. 86; *Opera* (Venice: Junctas 1562), fol. 189$^v$-190$^r$.

2. *Physica* VI.1 231$^b$9–10; *Aristoteles Latinus* VII.1, fasc. 2, ed. Fernand Bossier and Jozef Brams (Leiden-New York: E. J. Brill 1990), 217.

3. *Physica* II.2 194$^a$8–9; *Aristoteles Latinus* VII.1, fasc. 2, p. 51.

4. *Meteorologia* III.2–5 371$^b$19–377$^a$27; Aquinas, *In Aristotelis libros De Caelo et mundo, De generatione et corruptione, Meteorologicorum Expositio,* ed. R. M. Spiazzi (Turin-Rome: Marietti 1952), Textus Aristotelis, 617–618, 621, 624–625, 628–629, 636–637.

5. Landulfus Caracciolo, *Super I Sententiarum,* Prologus, q. 2. a. 4; Vienna, Nationalbibliothek 1496, fol. 5$^{rb}$.

6. *Analytica Posteriora* I.6 75$^a$18–20; *Aristoteles Latinus* IV.1–4, ed. L. Minio-Paluello and B. G. Dod (Bruges-Paris: De Brouwer 1968), 18–19.

7. *Analytica Posteriora* I.13 78$^b$39; *Aristoteles Latinus* IV.1–4, p. 32.

8. *Analytica Posteriora* I.13 78$^b$34; *Aristoteles Latinus* IV.1–4, p. 31.

9. Thomas Aquinas, *In libros Posteriorum Analyticorum Expositio* I, lect. 25, no. 208; ed. R. Spiazzi (Turin: Marietti 1964), 231.

10. Thomas Aquinas, *In libros Posteriorum Analyticorum Expositio* I, lect. 15, no. 209; *ed. cit.* p. 231.

11. *Analytica Posteriora* I.13 79$^a$14–17; *Aristoteles Latinus* IV.1–4, p. 32.

12. *Analytica Posteriora* I.13 79$^a$7–10; *Aristoteles Latinus* IV.1–4, p. 32.

13. *Physica* II.2 194$^a$8–11; *Aristoteles Latinus* VII.1, fasc. 2, p. 51.

14. Thomas Aquinas, *In VIII libros Physicorum Aristotelis Expositio* L. II, l. iii, no. 164; p. 84b.

15. Walter Burley, *Expositio in libros VIII de physico auditu* (Venice: Johannes Herbort de Almania 1482), fol. F$_8$$^{ra-b}$.

16. Egidius Romanus, *In libros de physico auditu Aristotelis Commentaria* L. II, l. 3; (Venice: Bonetus Locatellus 1502), fol. 29$^{ra}$.

17. Thomas Aquinas, *In libros Posteriorum Analyticorum Expositio* I, lect. 41, no. 361–365; *ed. cit.* pp. 300–301.

18. Egidius Romanus, *In libros de physico auditu Aristotelis Commentaria* L. II, l. 3; *ed. cit.*fol. 28$^{va}$.

19. Thomas Aquinas, *In VIII libros Physicorum Aristotelis Expositio* L. II, l. iii, no. 161–162; *ed. cit.* p. 83.

20. Paulus Venetus, *Expositio in libros Posteriorum Aristotelis* (Venice: Simon Papiensis 1494), fol. 42$^{ra-b}$.

21. *Aristotelis Metaphysicorum libri XIII cum Averrois commentariis* L. XII, c. 1, § 2; *Opera* (Venice: Junctas 1562; reprt. Frankfurt am Main: Minerva 1962), fol. 291$^{rb}$E-291$^{vb}$L.

22. *Physica* II.2 193$^b$32–194$^a$10; *Aristoteles Latinus* VII.1, fasc. 2, pp. 50–51.

23. *Forsitan Metaphysica* II.1 993$^b$30–31; *Aristoteles Latinus* XXV.2, ed. G. Vuillemin-Diem (Leiden: Brill 1976), 37.

24. *Topica* VI.4 141ᵃ35–38; *Aristoteles Latinus* V.1–3, ed. L. Minio-Paluello (Bruxelles-Paris: De Brouwer 1969), 119.

25. *Metaphysica* VII.12 1037ᵇ21–26; *Aristoteles Latinus* XXV.2, p. 146.

26. Antonius de Carlenis de Napoli, *Quaestiones in libros I–II Analyticorum Posteriorum Aristotelis* I, q. 16; fol. 17ʳᵃ-18ʳᵇ, esp. 17ᵛᵃ⁻ᵇ.

27. *Aristotelis Metaphysicorum libri XIII cum Averrois commentariis* L. VI, sum. lib., § 1; *Opera* fol. 144ʳᵇD-144ᵛᵃI.

28. Egidius Romanus, *In libros de physico auditu Aristotelis commentaria* Prol., pars II, dub. 2; *ed. cit.* fol. 2ᵛᵃ⁻ᵇ.

29. Thomas Aquinas, *In VIII libros Physicorum Expositio* L. I, l. i, no. 3; *ed. cit.* p. 36.

30. Cf. Aquinas, *In VIII libros Physicorum Expositio* L. II, l. xi, no. 243; *ed. cit* p. 118a.

31. Thomas Aquinas, *Summa theologiae* I, i, a. 2; ed. P. Caramello (Turin: Marietti 1952), vol. 1, p. 3.

32. Duns Scotus, *Reportata Parisiensia* Prol., q. 2; *Opera omnia* vol. 22 (Paris: Vivès 1894), 34b–37a.

33. Egidius Romanus, *Primus Sententiarum* 2 princ., q. 1, a. 2; (Venice: Octavianus Scotus 1521), fol. 4ᵛᵃK-L.

34. Antonius de Carlenis de Napoli, *Liber questionum super IV libros Sententiarum* Prol. q. 2; edn. pp. 20–27.

35. *Analytica Posteriora* I.13 78ᵇ32ff.; *Aristoteles Latinus* IV.1–4, pp. 31–32.

36. *Forsitan* tertium notandum; *supra*, p. 47.

37. *Supra*, p. 46.

38. *Analytica Posteriora* I. 13 79ᵃ14; *Aristoteles Latinus* IV.1–4, p. 32.

39. Egidius Romanus, *Expositio in libros Posteriorum* (Venice: Bonetus Locatellus 1488), fol. F₇ʳᵃ.

40. Paulus Venetus, *Expositio in libros Posteriorum Aristotelis; ed. cit.* fol. 42ʳᵇ.

41. *Physica* I.2 184ᵇ27–185ᵃ20; *Aristoteles Latinus* VII.2, ed. Fernand Bossier and Jozef Brams (Leiden-New York: E. J. Brill 1990), 4. Cf. alia transl. *Aristoteles Latinus* VII.1, fasc. 2, pp. 9–10.

42. Ibid.

43. *Supra*, p. 45. Landulfus, *Super I Sententiarum* Prol., q. 2, a. 4; Vienna, Nationalbibliothek 1496, fol. 5ʳᵇ.

44. *Supra*, p. 47.

45. Antonius de Carlenis de Napoli, *Quaestiones in libros I–II Analyticorum Posteriorum Aristotelis* I, q. 16; Rome, B. Casanatense 1025, fol. 17ʳᵃ-18ʳᵇ, esp. 17ᵛᵇ-18ʳᵃ.

46. Egidius Romanus, *Expositio in libros Posteriorum* fol. F₇ʳᵃ.

47. *Analytica Posteriora* I.13 78ᵇ38; *Aristoteles Latinus* IV.1–4, p. 31.1. *Aristotelis Posteriorum Resolutoriorum libri duo cum Averrois . . . magnis commentariis. . . .* I, comm. 86; *Opera* (Venice: Junctas 1562), fol. 189ᵛ-190ʳ.

# APPENDIX 1

## Description of Oxford, Bodleian Library,
## Canon. misc. 573

14th–15th C. Composite volume in two sections. Leather binding, spine tooled in red and gold. i + 377 + i.

I. Vellum and paper, 29.6 × 20.5 cm., ff. 171, double columns of 57–62 lines, written space 22.5 x 15.3 cm. Prague, 1384–1385.

Collation: i–vi$^{16}$ vii$^{14}$ vii–x$^{16}$ xi$^{16}$ (wants 3). With the exception of first two quires, catchwords in lower right margin.

2° fol.: cunctorum

**Conrad of Ebrach,** *Opus super quatuor libros Sententiarum.*

fol. 1ra: <Bk. I> Flumen Dei repletum est aquis Psalmo xl sexto. Spiritali dulcedine celestium fluentorum inebriata. . . .

fol. 32va: . . . et sic est finis istius questionis. etc.

fol. 33ra: Utrum hec consequentia est bona: Spiritus Sanctus non procedit. . . .

fol. 53va: . . . et sic est finis istius questionis et per consequens questionum totius libri Sententiarum primi. Per fratrem Johannem de Reiz Australem nacione qui eas finivit Prague sabato in vigilia sancti Jacobi apostoli anno domini MCCCLXXXIIII Dominus sit benedictus. Amen dicant omnia etc.

fol. 54ra: <Bk. II> Flumen Dei repletum est aquis Psalmo lx sexto. Fons sapientie verbum Dei inexpressis. . . .

fol. 93vb: . . . et sic est finis questionum Sententiarum libri secundi qui est finitus anno domini MCCCLXXXIIII feria sexta in die sancte ludmille martiris et vidue in terra boemie qui dies celebratur in sequenti die octave nativitatis sancte marie.

fol. 94ra: <Bk. III> Flumen dei repletum est aquis Psalmo lx quarto. Dulcissimum sancti spirite organum. . . .

fol. 108va: . . . et sic est finis questionum tertii libri Sententiarum anno domini MCCCLXXXIIII feria quarta finite in octava sancti martiris Wenheslay, Boemorum ducis. Deo laus in eternum. Amen.

fol. 109ra: <Bk. IV> Flumen Dei repletum est aquis Psalmo sexagesimo quarta. Doctor melliflus venerabilis. . . .

fol. 156rb: . . . et sic est finis libri quarti Sententiarum reverendi magistri Conradi de Ebraco ordinis sancti Bernardi Clarevallis abbatis.

fol. 156va: Questio in vesperiis. Utrum latitudo in cuiuslibet culpe mortalis ymaginalis sit mensuranda penes discessum voluntatis. . . .

fol. 162rb: Explicit opus questionum super quatuor libros Sententiarum reverendi magistri Conradi de Ebraco, ordinis Cystersiensium. Scriptum Prage in conventu sancti Thome per manus fratris Johannis de Reiz, ordinis fratrum heremitarum sancti Augustini, pro tunc ibidem studentis, sub anno Domini millesimo tricentesimo octuagesimo quinto, feria sexta infra octavas Penthecostes; Deo laus in eternum. Amen. Qui te furetur in furca vita privetur.

Iste liber est monasterii sancte Marie de Caritate Venetiarum.

fol. 162va: Index prohemium.

fol. 163r: "Statio huius libri est in septima sede."

fol. 163v-164r blank.

fol. 164va-165ra: Index questionum.

fol. 165rb-171r blank.

II. Paper, 29.4 x 21.0 cm., ff. 206, double columns of 34–37 lines, written space 20.2 x 13.7 cm. Italian hand(s?) s. XV. Three watermarks throughout: licorne of the Italian type (similar to Briquet nn. 9957, 9960, 9971), té (Briquet n. 14089), unidentified circle design.

Collation: i–iv$^{12}$ v$^{12}$ (wants one) vi–viii$^{12}$ ix$^8$ x–xi$^{12}$ xii$^6$ xiii$^8$ xiv–xviii$^{12}$ xix$^{10}$ (wants 4). Catchwords in center lower margin of each quire.

2° fol. declaratum

**Antonius de Carlenis,** *Liber questionum super quatuor libros Sententiarum.*

fol. 172ra: Defective at beginning. Inc: . . . dicit theologiam esse scientiam subalternam. . . .

fol. 271va: . . . predestinatus enim erat quod partibus(?) gregorii(?) salvaretur. Sequitur secundus liber Sententiarum.

fol. 272–273 blank.

fol. 274ra: Secundus liber. Circa primam divisionem secundi libri Sententiarum, queritur utrum mundi productio extrinsica de necessitate presupponat productiones intrinsicas in divinis. . . .

fol. 310vb: . . . ut dictum est in responsione ad 12m.

fol. 311 blank.

fol. 312ra: Tertius liber Sententiarum. Queritur utrum tres persones potuerunt accipere unam naturam. . . .

fol. 347ra: . . . secundum seminalem rationem in Adam. Laus Deo. Amen.

fol. 247v blank.

fol. 348ra: <Bk. IV> Queritur utrum in sacramento sit aliqua Christus creativa gratie. . . .

fol. 375rb: . . . qualis sit sanctorum gloria. Deo gratia. Amen.

fol. 375va: <Index questionum.> Super primum, et primo circa prohemium. Queritur utrum doctrina sacra sit scientia. . . .

fol. 377ra: Explicit liber super quatuor libros Sententiarum compositus a reverendissimo domino Domno A. archiepiscopo Amalphitano, sacre theologie magistro egregio, ordinis Praedicatorum dignissimo, etc. Amen.

fol. 377v blank.

# APPENDIX 2

## Variant Incipit
### in the
### Questiones in IV libros Sententiarum
### Oxford, Bodleian Library, Canon. misc. 573, fol. 172ra.

### Prologue, Q. 1, primum notandum

'. . . dicit* theologiam esse scientiam subalternam, non intelligit quod theologia sit scientia subalterna, <quantum ad hoc, quod scientia subalterna> inventa ab homine habet processum scientificum, sed quantum ad hoc, quod habet similitudinem cum ea in hoc, quod sicut scientia subalterna proprie dicta inventa ab homine habet sua principia saltem quantum ad propter quid <est> ut credita, ita etiam theologia habet principia sua credita. Sed non solum quantum ad† propter quid, sed etiam quantum ad quia est. Et dicit quod scientia subalterna, nisi coniungatur scientie subalternanti in eodem subiecto, non habet proprie rationem‡ scientie, et hoc solum intelligendum est quantum ad propter quid <est> suorum principiorum; theologia autem, quantum ad propter quid est et quantum ad quia est, habet principia credita. Et ideo non habet rationem scientia proprie dicte nec quantum ad propter quid est nec quantum ad quia est. Unde etiam frater Thomas in I *Sententiarum* videtur dicere quod theologia non proprie sed largo modo dicta scientia subalterna.'[1] Hec omnia. Dicit hec ad declarationem dictorum sancti doctoris ad fratrem Hownem(?) magistri Almerici[2] generalis ordinis predicatorum.

---

*O fol. 172ra.
†quantum ad *transp.* O.
‡rationem *corr.* O.

## Notes

1. Herve Natalis, *Defensa doctrinae sancti Thomae* prima pars, II, 1., a. 7; ed. Engelbert Krebs, *Theologie und Wissenschaft nach der Lehre der Hochscholastik* (Münster i. W.: Aschendorffsche Verlagsbuchhandlung 1912), 37*.

2. Aimericus de Placentia, magister generalis Ordinis Praedicatorum (1304–1311).

# APPENDIX 3

**Tabula questionum**
**Antonius de Carlenis,**
*Questiones in libros*
*I–II Analyticorum Posteriorum Aristotelis*
**Chicago, Newberry Library, Case MS 97,5**

L. I

1. Queritur primo utrum subiectum libri *Posteriorum* sit sillogismus demonstrativus. Et arguitur primo quod non. (fol. 2ra)

2. Secundo queritur utrum omnis doctrina et omnis disciplina intellectiva ex preexistenti fiat cognitione. Arguitur quod non. . . . (fol. 3vb)

3. Tertio queritur utrum sint tantum due precognitiones et tria precognita. Et arguitur primo quod non. (fol. 4rb)

4. Quarto queritur utrum maior et minor precognoscantur ante conclusionem. Arguitur primo quod non. (fol. 5rb)

5. Quinto queritur utrum scire sit bene diffinitum, scilicet quod est cognoscere rem per causam et quantum eius est causa et quod non contingat aliter se habere. Arguitur quod non. (fol. 6ra)

6. Sexto queritur utrum diffinitio demonstrationis que dicitur quod ex primis veris et immediatis prioribus et notioribus causis conclusionis sit bene assignata. Arguitur quod non. (fol. 6vb)

7. Septimo queritur utrum omnis demonstratio sit de ente. Probatur quod non. (fol. 8ra)

8. Octavo queritur utrum demonstratio sit ex immediatis, et arguitur quod non. (fol. 8vb)

9. Nono queritur utrum necesse sit magis adherere principiis quam conclusionibus. Arguitur quod non. (fol. 9vb)

10. Decimo queritur utrum propter quid unumquodque tale et illud magis est. Arguitur quod non. (fol. 10va)

11. Undecimo queritur utrum detur demonstratio circularis. Arguitur quod sic. (fol. 11vb)

12. Duodecimo queritur utrum dici de omni bene hic diffiniatur, scilicet quod requirit universalitatem suppositorum et temporis, et arguitur primo quod non. (fol. 12vb)

13. Tertio decimo queritur utrum bene assignentur duo primi modi perseitatis. Arguitur quod non. (fol. 13va)

14. Quarto decimo queritur utrum tertius modus et quartus per se bene a Philosopho ponantur. Arguitur primo quod non. (fol. 15ra)

15. Quinto decimo queritur utrum diffinitio universalis sit ab Aristotele bene assignata, cum dicitur quod universale est quod cum sit de omni per se et secundum quod ipsum est. Arguitur primo quod non. (fol. 16vb)

16. Sexto decimo queritur utrum demonstratio sit semper ex necessariis et de necessariis. Arguitur primo quod non. (fol. 18ra)

17. Decimo septimo queritur utrum demonstratio possit esse ex extraneis, et arguitur quod sic. (fol. 19rb)

18. Decimo octavo queritur utrum bene assignentur principia complexa demonstrationis, videlicet dignitas, suppositio et diffinitio. Arguitur quod non. (fol. 20rb)

19. Decimo nono queritur utrum primum principium non contingit simul affirmare et negare, ingrediatur formaliter demonstrationem, et arguitur quod sic. (fol. 21ra)

20. Vicesimo queritur utrum scientie mathematice sive doctrinales sint certissime. Arguitur quod non. (fol. 22ra)

21. Vicesimo primo queritur utrum convenienter dividatur demonstratio in demonstrationem propter quid et quia, et arguitur quod non. (fol. 23rb-va)

22. Vicesimo secundo queritur utrum convenienter assignetur ab Aristotele distinctio scientie subalternantis et subalternate. Et arguitur quod non. (fol. 24vb)

23. Vicesimo tertio queritur utrum deficiente sensu fit necessarie deficienter scientiam que per illum sensum habetur. Arguitur quod non. (fol. 26rb)

24. Vicesimo quarto queritur utrum propositio in qua removetur unum predicamentum ab altero sit immediata negative. Arguitur quod non. (fol. 27rb)

25. Vicesimo quinto queritur utrum inter determinata extrema possuit esse media infinita. Arguitur quod sic. (fol. 28rb)

26. Vicesimo sexto queritur utrum demonstratio universalis particulari sit potior, et arguitur breviter quod non. (fol. 29rb)

27. Vicesimo septimo queritur utrum demonstratio affirmativa sit potior quam negativa, et ostensiva sit potior quam ducens ad impossibile. Arguitur quod non. (fol. 30ra)

28. Vicesimo octavo queritur utrum demonstratio ostensiva sit potior illa que est ad impossibile. Arguitur quod non. (fol. 30vb)

29. Vicesimo nono queritur utrum unitas scientie capiatur ex unitate subiecti. Arguitur quod non. (fol. 31rb)

30. Tricesimo queritur utrum scibile et scientia differant ab opinabili et opinione. Arguitur quod non. (fol. 32vb)

L. II

1. Circa secundum librum *Posteriorum* primo queritur utrum questiones sint tantum quattuor equales numero hiis que vere scimus. Arguitur quod non. (fol. 33va)

2. Secundo queritur utrum omnis questio sit medii. Arguitur quod non. (fol. 34rb)

3. Tertio queritur utrum unius et eiusdem possit esse diffinitio et demonstratio. Arguitur quod sic. (fol. 34vb)

4. Quarto queritur utrum diffinitio possit demonstrari de suo diffinito in eo quod diffinitur, et primo quod sic. (fol. 35rb)

5. Quinto queritur utrum diffinitio possit demonstrari diffinito aliquo sillogismo reduplicativo, quia tales sunt conclusiones, quales sunt premisses. . . . (fol. 35vb)

6. Sexto queritur utrum possit demonstrari per sillogismum divisivum sine per sillogismum diffinitivum. Arguitur quod sic. (fol. 36vb)

7. Septimo queritur utrum solummodo demonstratio declaret esse. Arguitur quod non. (fol. 37va)

8. Octavo queritur utrum medium in demonstratione potissima sit diffinitio subiecti. Arguitur quod non. (fol. 38vb)

9. Nono queritur utrum quodlibet genus cause possit fieri demonstratio respectu effectus. Arguitur quod non. (fol. 43rb)

10. Decimo queritur utrum per viam demonstrationis generis indifferentis possimus venari ipsum quod quid est sive diffinitionem. Arguitur quod non. (fol. 44rb)

11. Undecimo queritur utrum plurium conclusionum possit esse idem medium. Arguitur quod non. (fol. 46ra)

12. Duodecimo queritur utrum cognitio primorum principiorum insit nobis a natura. Arguitur quod sic. (fol. 46rb)

# BIBLIOGRAPHY

## 1. Manuscripts Cited

Cambridge, Gonville and Caius College, MS 370(592)
Chicago, Newberry Library, Case MS 97,5
Florence, Biblioteca Nazionale Centrale, B.VI.340.1
Milan, Biblioteca Trivulziana 1682 (460)
Naples, Biblioteca Nazionale VIII.G.75
Oxford, Bodleian Library, Canon. misc. 573
Rome, Biblioteca Casanatense 1025
Vatican, Biblioteca Reginensis lat. 392
Vatican, Borgh. lat. 27
Vatican, *Registra Vaticana* 388
Vatican, Vat. lat. 817
Vatican, Vat. lat. 3964
Vatican, Vat. lat. 7135–7136
Vienna, Nationalbibliothek 1496

## 2. Printed Sources

*Acta Capitulorum Generalium Ordinis Praedicatorum*, ed. B. M. Reichert. 9 volumes. [Monumenta Ordinis Fratrum Praedicatorum Historica 3, 4, 8–14]. Rome: Typographia Polyglotta, 1898–1904.

Aegidius Romanus. *In libros de physico auditu Aristotelis Commentaria.* Venice: Bonetus Locatellus, 1502.

Alce, P. V., and P. A. D'Amato. *La Biblioteca di S. Domenico in Bologna.* Firenze: Leo S. Olschki, 1966.

Altamura, Ambrosius de. *Bibliothecae Dominicanae . . . .* Rome: Nicolai Tinassii, 1677.

Ariew, Roger. *Medieval Cosmology: Theories of Infinity, Place, Time, Void and the Plurality of Worlds.* Chicago: University of Chicago Press, 1985.

Aristotle. *Analytica Posteriora; Aristoteles Latinus* IV.1–4, ed. L. Minio-Paluello and B. G. Dod. Bruges-Paris: De Brouwer, 1968.

———. *Metaphysica; Aristoteles Latinus* XXV, ed. G. Vuillemin-Diem. Leiden: Brill, 1976.

———. *Physica; Aristoteles Latinus* VII.1–2, ed. Fernand Bossier and Jozef Brams. Leiden-New York: Brill, 1990.

———. *De sophisticis elenchis; Aristoteles Latinus* VI, ed. B. G. Dod. Leiden-Bruxelles: Brill, De Brouwer, 1975.

———. *Topica; Aristoteles Latinus* V.1–3, ed. L. Minio-Paluello. Bruxelles-Paris: De Brouwer, 1969.

Augustine. *De Trinitate*, ed. W. J. Mountain, *Corpus Christianorum*, ser. lat. 50A. Turnholt: Brepols, 1968.

Averroes. *Aristotelis Metaphysicorum libri XIII cum Averrois commentariis; Opera*, volume 8. Venice: Junctas, 1562; reprt. Frankfurt am Main: Minerva, 1962.

———. *Aristotelis Posteriorum Resolutoriorum libri duo cum Averrois . . . magnis commentariis. . . . ; Opera*, volume 1, part 2. Venice: Junctas, 1562; reprt. Frankfurt am Main: Minerva, 1962.

Bertolà, M. *I due primi Registri di prestito della Bibl. Apost. Vaticana*. Vatican: B. Apostolica Vaticana, 1942.

*Bibliorum sacrorum iuxta Vulgatam Clementinam. Editio nova*, ed. Aloisius Gramatica. Rome: Typis Polyglottis Vaticanis, 1959.

Bignami Odier, Jeanne. *La Bibliothèque Vaticane de Sixte IV à Pie XI* [Studi e testi 272]. Vatican: B. Apostolica Vaticana, 1973.

Briquet, C. M. *Les Filigranes: Dictionnaire historique des marques du papier dès leur apparition vers 1282 jusqu'en 1600*. 4 volumes. 2nd edn. Leipzig: Karl von Hiersemann, 1923.

*Chartularium studii Bononiensis*. Bologna: Presso el Commissione per la storia dell'Università di Bologna, 1909– .

*Chartularium Universitatis Parisiensis*, ed. H. Denifle and E. Chatelain. 4 volumes. Paris: ex typis fratrum Delalain, 1889–1897.

Clagett, Marshall. *Giovanni Marliani and Late Medieval Physics*. New York: Columbia University Press, 1941.

Coxe, H. O. *Catalogi codicum manuscriptorum Bibl. Bodleianae*, Pars III. Oxford: Clarendon Press, 1854.

Dionysius Cartusiensis. *Commentaria in IV libros Sententiarum; Opera omnia* vol. 19. Tournai: Typis Cartusiae S. M. de Pratis, 1902.

Dominicus de Flandria. *Quaestiones perutiles in D. Thomae Aquinatis Commentaria super libris Posteriorum Analyticorum Aristotelis*. Venice: Apud Michaelem Berniam Bibliopolam Bononiensem, 1587.

Dondaine, A. "Abbréviations latines et signes recommandés pour l'apparat critique des éditions de textes médiévaux," *Bulletin de la Société internationale pour l'étude de la philosophie médiévale* 2 (1960): 142–149.

———. "Variantes de l'apparat critique dans les éditions de textes latins médiévaux," *Bulletin de la Société internationale pour l'étude de la philosophie médiévale* 4 (1962): 82–100.

Duhem, Pierre. *Le Système du monde*. 10 volumes. Paris: Hermann, 1913–1959.

Durand de St. Pourçain. *In Petri Lombardi Sententias Theologicas Commentariorum libri IIII*. Venice: Typographia Gerraea, 1571; reprt. Ridgewood, NJ: Gregg Press, 1964.

Egidius Romanus. *Expositio in libros Posteriorum*. Venice: Bonetus Locatellus, 1488.

———. *In libros de physico auditu Aristotelis Commentaria*. Venice: Octavianus Scotus, 1502.

———. *Primus Sententiarum*. Venice: Octavianus Scotus, 1521.

Emery, Jr., Kent. "Theology as a Science: The Teaching of Denys of Ryckel (Dionysius Cartusiensis, 1402–1471)," *Knowledge and Science in Medieval Philosophy. The Proceedings of the Eighth International Congress of Medieval Philosophy (SIEPM), Helsinki, 24–29 August 1987*, ed. R. Työrinoja, A. I. Lehtinen, D. Føllesdal. Helsinki: Société Philosophique de Finlande, 1990. Volume 3, pp. 376–388.

Fontana, Vincentio Maria. *Sacrum theatrum dominicanum*. Rome: Nicolai Angeli Tinassii, 1665.

Franciscus de Mayronis. *In IV libros Sententiarum*. Venice: O. Scotus, 1520.

Gilbert, Neal. *Renaissance Concepts of Method*. New York: Columbia University Press, 1960.

Glorieux, Palémon. *La faculté des arts et ses maîtres au XIIIᵉ siècle*. Paris: Vrin, 1971.

———. *La littérature quodlibétique de 1260 à 1320*. 2 volumes. Le Saulchoir: Kain, 1925; Paris: Vrin, 1935.

———. *Répertoire des maîtres en théologie de Paris au XIIIᵉ siècle*. 2 volumes. Paris: Vrin, 1933–34.

Godofridus de Fontibus. *Quodlibetum VIII; Les Philosophes Belges. Textes et études 4. Le huitième quodlibet de Godefroid de Fontaines*, ed. J. Hoffmans. Louvain: Institut supérieur de philosophie de l'Université, 1924.

Goldstein, Thomas. *Dawn of Modern Science*. Boston: Houghton Mifflin, 1980.

Grassi, Onorato. "La questione della teologia come scienza in Gregorio da Rimini," *Rivista di filosofia neo-scolastica* 58 (1976): 610–644.

Gregory of Rimini. *Lectura super primum et secundum Sententiarum*, ed. A. Damasus Trapp and Venicio Marcolino. 6 volumes. Berlin-New York: De Gruyter, 1981– .

Hay, Denys. *The Church in Italy in the Fifteenth Century*. Cambridge: Cambridge University Press, 1977.

Henricus Gandavensis. *Summae quaestionum ordinariarum*. Paris: J. Badius, 1520; reprt. St. Bonaventure, NY: Franciscan Institute, 1953.

Hervé Natalis. *Defensa doctrinae S. Thomae*, ed. Engelbert Krebs, *Theologie und Wissenschaft nach der Lehre der Hochscholastik*. Münster i. W.: Aschendorffsche Verlagsbuchhandlung, 1912.

———. *In IV libros Sententiarum*. Paris: Dyonisius Moreau, 1647.

Jerome. *Epistola LIII. ad Paulinum*, ed. Isidorus Hilberg. CSEL 54. Vienna-Leipzig: F. Tempsky, G. Freytag, 1910.

Joannes de Neapoli. *Questiones disputatae*. Neapoli: C. Vitalis, 1618; reprt. Ridgewood, NJ: Gregg Press, 1966.

Johannes Capreoli. *Defensiones theologiae divi Thomae Aquinatis*, ed. C. Paban and T. Pègues. 7 volumes. Tours: Alfred Cattier, 1900–1908; reprt. Frankfurt/Main: Minerva, 1967.

Johannes Duns Scotus. *Ordinatio*, Prologue, ed. P. Perantoni, *Opera Omnia*, vol. 1. Vatican: Polyglottis, 1950.

———. *Quaestiones in librum tertium Sententiarum; Opera omnia* vol. 15. Paris: Vivès, 1894.

————. *Reportata Parisiensia; Opera omnia* vols. 22–23. Paris: Vivès, 1894.

Kaeppeli, Tommaso. "Antiche biblioteche Dominicane in Italia," *Archivum fratrum Praedicatorum* 36 (1955): 5–80.

————. "Dalle pergamene di S. Domenico di Napoli," *Archivum Fratrum Praedicatorum* 32 (1962): 285–326.

————. *Scriptores ordinis Praedicatorum Medii Aevi.* 3 volumes. Rome: S. Sabina, 1970– .

Kristeller, Paul Oskar. "Thomism and the Italian Thought of the Renaissance," in *Medieval Aspects of Renaissance Learning,* ed. and trans. Edward P. Mahoney. Durham, N.C.: Duke University Press, 1974. Pp. 29–91.

Kuttner, Stephan. "Notes on the Presentation of Text and Apparatus in Editing Works of the Decretists and Decretalists," *Traditio* 15 (1959): 452–464.

Laird, Walter Roy. "The *Scientiae Mediae* in Medieval Commentaries on Aristotle's *Posterior Analytics.*" Unpublished PhD dissertation, University of Toronto, Centre for Medieval Studies, 1983.

Laurent, M.-H. *Fabio Vigili et les bibliothèques de Bologne au début du XVIᵉ siècle* [Studi e testi 105]. Vatican: Biblioteca Apostolica Vaticana, 1943.

Livesey, Steven J. "*Metabasis:* The Interrelationship of the Sciences in Antiquity and the Middle Ages." Unpublished PhD dissertation, The University of California, Los Angeles, 1982.

————. "On Pierre Duhem," *Science in Context* 1 (1987): 363–370.

————. "The Oxford Calculatores, Quantification of Qualities, and Aristotle's Prohibition of *metabasis,*" *Vivarium* 24 (1986): 50–69.

————. "Science and Theology in the Fourteenth Century: the Subalternate Sciences in Oxford Commentaries on the *Sentences,*" *Synthèse* 83 (1990): 273–292.

————. *Theology and Science in the Fourteenth Century. Three Questions on the Unity and Subalternation of the Sciences from John of Reading's Commentary on the Sentences.* Leiden: E. J. Brill, 1989.

————. "William of Ockham, the Subalternate Sciences, and Aristotle's Prohibition of *metabasis,*" *British Journal for the History of Science* 18 (1985): 127–145.

McKirahan, Richard D. "Aristotle's Subordinate Sciences," *British Journal for the History of Science* 11 (1978): 197–220.

Meersseman, G. "Antonius de Carlenis O.P., Erzbischof von Amalfi," *Archivum fratrum Praedicatorum* 3 (1933): 81–131.

————. "Een Vlaamsch Wijsgeer: Dominicus van Vlaanderen," *Thomistisch Tijdschrift* 1 (1930): 385–400, 590–592.

————. "Ergaenzung zum Schrifttum des Antonius de Carlenis von Neapel O.P.," *Archivum Fratrum Praedicatorum* 5 (1935): 357–363.

Origlia, Gianguiseppe. *Istoria dello Studio di Napoli.* Naples: Giovanni di Simone, 1753.

Paulus Venetus. *Expositio in libros Posteriorum Aristotelis.* Venice: Simon Papiensis, 1494.

Petrus Aureoli. *Scriptum super I Sententiarum*, ed. Eligius M. Buytaert. 2 volumes. St. Bonaventure, NY: Franciscan Institute, 1952–1956.

Piana, Celestino. "La facoltà teologica dell' Università di Bologna nel 1444–1458," *Archivum Franciscanum Historicum* 53 (1960): 361–441.

――――. *Nuove ricerche su le Università di Bologna e di Parma*. Quaracchi: Collegium S. Bonaventurae, 1966.

Quétif, Jacobus and Jacobus Echard. *Scriptores Ordinis Praedicatorum Recensiti . . . .* 2 volumes. Paris: Ballard-Simart, 1719.

Randall, John Herman. "Paduan Aristotelianism Reconsidered," in *Philosophy and Humanism. Renaissance Essays in Honor of Paul Oskar Kristeller*, ed. E. P. Mahoney. New York: Columbia University Press, 1976. Pp. 275–282.

――――. *The School of Padua and the Emergence of Modern Science*. Padua: Editrice Antenore, 1961.

*I Rotuli dei lettori legisti e artisti dello studio Bolognese dal 1384 al 1799*, ed. Umberto Dallari. 4 volumes. Bologna: Fratelli Merlani, 1888–1924.

Saenger, Paul. *A Catalogue of the Pre-1500 Western Manuscript Books at the Newberry Library*. Chicago: University of Chicago Press, 1989.

Santoro, Caterina. *I codici medioevali della Biblioteca Trivulziana*. Milano: Biblioteca Trivulziana, 1965.

Schikowski, Ulrich. "Dominicus de Flandria O.P. († 1479), sein Leben, seine Schriften, seine Bedeutung," *Archivum Fratrum Praedicatorum* 10 (1940): 169–221.

Schmitt, Charles B. *A Critical Survey and Bibliography of Studies on Renaissance Aristotelianism 1958–1969*. Padua: Ed. Antenore, 1971.

――――. *Aristotle in the Renaissance*. Cambridge, MA: Harvard University Press, 1983.

――――. "Towards a Reassessment of Renaissance Aristotelianism," *History of Science* 11 (1973): 159–193.

Tachau, Katherine. *Vision and Certitude in the Age of Ockham. Optics, Epistemology and the Foundations of Semantics 1250–1345*. Leiden: E. J. Brill, 1988.

Thomas Aquinas. *In Aristotelis libros De Caelo et mundo, De generatione et corruptione, Meteorologicorum Expositio*, ed. R. M. Spiazzi. Turin-Rome: Marietti, 1952.

――――. *In libros Posteriorum Analyticorum Expositio*, ed. R. Spiazzi. Turin: Marietti, 1964.

――――. *In VIII libros Physicorum Aristotelis Expositio*, ed. P. M. Maggiòlo. Turin-Rome: Marietti, 1954.

――――. *Scriptum super libros Sententiarum*, (L. I-II), ed. P. Mandonnet. 2 volumes. Paris: P. Lethielleux, 1910.

――――. *Summa theologiae*, ed. P. Caramello. 4 volumes. Turin: Marietti, 1952.

――――. *De veritate; Quaestiones disputatae* volume 1, ed. R. Spiazzi. Turin-Rome: Marietti, 1964.

Thorndike, Lynn. *A History of Magic and Experimental Science.* 8 volumes. New York: Columbia University Press, 1923–1958.

———. *Science and Thought in the Fifteenth Century.* New York: Columbia University Press, 1929.

Torraca, Francesco, *et al. Storia del Università di Napoli.* Naples: Riccardo Ricciardi Editore, 1924.

Tuszyńska, Franciszka. "Materiały do Stanu Badań nad Filozofią Scholastyczną XV Wieku," *Studia mediewistyczne* 2 (1961): 5–99.

Vasoli, Cesare. "La cultura dei secoli XIV–XVI," in *Atti del primo convegno internazionale di ricognizione delle fonti per la storia della scienza Italiana: I secoli XIV–XVI,* ed. Carlo Maccagni. Firenze: G. Barbèra, 1967. Pp. 31–105.

Xiberta, Bartholomaeus F. M. "De Summa Theologiae magistri Gerardi Bononiensis ex Ordine Carmelitarum," *Analecta ordinis Carmelitarum* ann. XIV, vol. 5 (1923): 3–54.

Wallace, William A. *Prelude to Galileo. Essays on Medieval and Sixteenth-Century Sources of Galileo's Thought.* Dordrecht-Boston: Reidel, 1981.

Walter Burley. *Expositio in libros VIII de physico auditu.* Venice: Johannes Herbort de Almania, 1482.

# INDEX

## Index to the Introduction

## II. Index to the Text

### 1. Auctores et Scripta

## 2. Doctrina

stereometria, 45, 52
subiectum: simpliciter et secundum quid,
    35; distinguitur per differentiam
    extraneam, 36; contractum, 44; *v.*
    passio, scientia, theologia

theologia: dicit habitum discursivum, 12;
    est habitus acquisitus, 13; notitia
    consequentialis, 5, 10, 11; procedit
    ex articulis fidei, 4, 7, 11, 13, 24; et
    sapientia 5; subiectum, 5, 22, 25–
    26. *v.* doctrina, habitus, scientia
  definitio: secundum Franciscum de
    Mayronis, 7; secundum Johannem
    de Neapoli 9–10

utrum sit scientia subalternata vel sub-
    alternans, 20–27; (non) est scientia
    subalternata scientie beatorum, 7,
    21, 49; est scientia (subalterna)
    largo modo dicta, 9, 10, 22, 24; non
    subalternatur methaphysice, 25, 45,
    51; non subalternat sibi alias huma-
    nas, 25

virtutes: infusae et acquisitae non sunt ad
    eundem finem (sec. Thomam), 13
'visio super assensam rectam est perfectis-
    sima', 21, 27, 45